# A Child Is Born

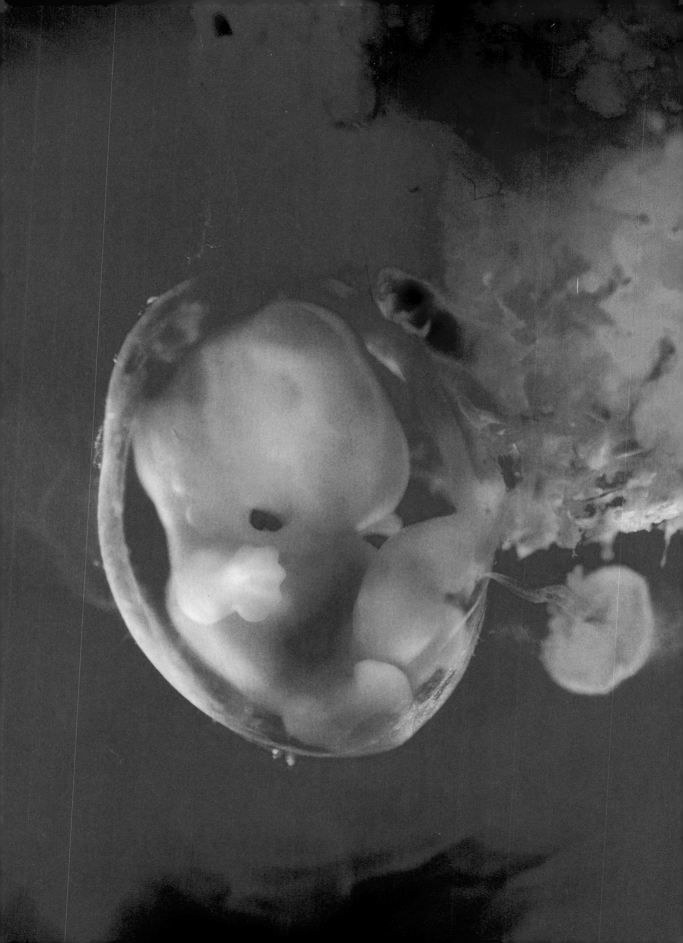

# A Child Is Born

Lennart Nilsson

*Text by* Lars Hamberger

Translated from the Swedish by Linda Schenck

A Merloyd Lawrence Book/Delacorte Press

# Contents

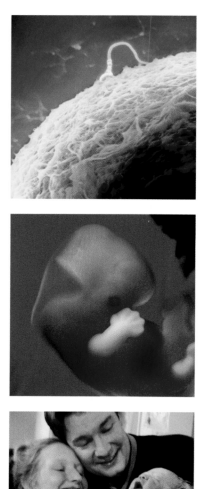

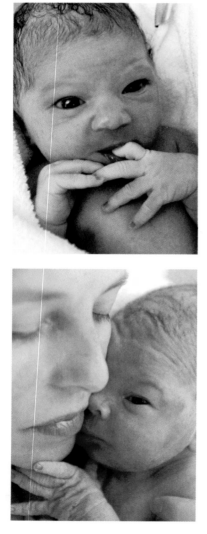

Today young couples planning to start a family ask more and more questions of themselves and of the society in which they live. Into what kind of world would we be bringing children? What sort of future can we offer them? What rights does a child have? The right to be born whole and healthy? To be welcome? To live with both a mother and a father? Sometimes they also have concerns about sterility, or fears of having a child with a deformity or a mental disability. Another key issue for both parents and legislators, who have to make difficult decisions, for instance, regarding abortion, is the question of when life begins. Many people hold the view that life begins at conception, at the moment intercourse results in pregnancy. Others consider life to begin when the fertilized egg becomes implanted in the woman's uterus, since without this contact the seed of life cannot develop into a human being. Others still are of the opinion that life cannot be said to begin until there are nerve cells conveying messages, such as pain or motion, or until a fetus can survive outside the mother's body.

In the developed parts of the world, it is becoming easier and safer to become pregnant and give birth to a baby. An expectant mother has access, throughout her pregnancy, to better help and more information and counseling today than ever before. She knows that she can personally contribute to reducing the risks for her child by eating the right foods, managing stress, and being careful in her use of medication, tobacco, and alcohol. Today the quality of ultrasound images is so good that it is possible for specialists to determine whether the fetus is healthy. The joy experienced by parents when they see their baby-to-be moving around in utero on the screen can have a positive effect on their entire experience of pregnancy. A number of other fetal diagnostic tests are available as well. Each woman is increasingly able to help plan the delivery itself. Babies born prematurely have a much better chance today to survive without problems, and in recent years we have also learned a great deal more about the causes of infertility and its treatment.

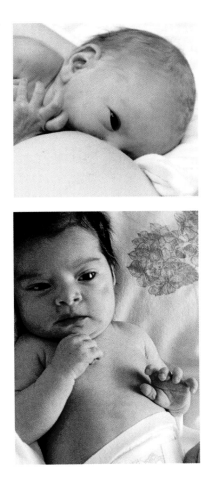

This book portrays the amazing miracle that takes place when a child is born. It has been our ambition to present as vivid and objective a picture as possible of the course of fertilization, pregnancy, and childbirth. We have tried to answer the commonest questions expectant parents tend to ask. Pregnancy is one of many normal phases of life—it is not an illness. Those closest to a pregnant woman and her caregivers can, in cooperation, help her experience pregnancy and childbirth as the most astonishing, impressive experience of her life and can give both parents an enduring sense of happiness over their privileged role in the development and birth of a child.

*Lars Hamberger*

# Woman and Man

**Love is an incredibly strong, enduring force and has been since time immemorial. The pattern is recognizable in every culture in our world: two people are mutually attracted and feel the irresistible urge to unite.**

## Six billion people

How many people can the world support? Six billion, ten billion, fifteen? No one can be sure. What we do know is that the population of the world is growing steadily. It has recently passed the six billion mark, a figure that has raised deep concern. But the picture is more complex than that. In the developing countries poverty is still rampant and the population is growing fast, while most industrialized countries actually have the opposite situation, with fertility rates at replacement levels or on the decline. In the long run this imbalance will result in migratory flows and population resettlements with unknown effects far into the future.

Every culture throughout history has been conscious of human reproduction, and a number of curious methods have traditionally been used to protect women from becoming pregnant, with counterparts in magical fertility rites. Not until recent centuries has mankind come to have a more rational, scientific attitude toward reproduction, pregnancy, and childbirth, and developments in the last thirty years have resulted in completely new ways of monitoring a pregnancy and guaranteeing a safe and sometimes even painless delivery. Couples who have trouble becoming pregnant also have access to better information and new techniques.

While contemporary society has left primitive beliefs behind, reproduction is still governed by complex political and religious systems with rules and regulations and sometimes even impenetrable taboos. Fortunately no social structure has ever managed to suppress love between women and men. And one key factor in this love is the desire of a couple to have children.

## Love at first sight

We are very sensitive to looks—for instance, when someone just stares at us for a few extra seconds—and even more so to the subtle body language of others. In the photo on the right, taken with a unique technique, the face that has made the person's heart beat just a bit faster appears on the retina of the observer.

## The decisive moment

When two sets of eyes meet suddenly and lock, time seems to come to a momentary halt in the middle of a busy street or, as the song has it, "across a crowded room." Many people recall this moment for the rest of their lives, and sometimes this impression remains our strongest image of a loved one. Our brains, our emotions, and our devotion may be blind to the aging process. When we first meet, we are receptive not only to the glint in someone's eye or to that person's appearance, but also to tone of voice and, perhaps most important, to body language and scent.

The male sex hormone testosterone and the female hormone estrogen contribute to the differences between the way men and women look, and they also affect each person's own particular scent via our pheromones, chemical substances that we release and for which each of us is uniquely coded.

## The essential sex hormones

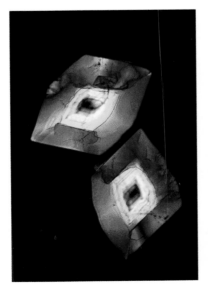

Male testosterone (right) and female estrogen (far right) transmit complex chemical messages. These hormones affect our appearance and our feelings and are essential to the reproduction process. Here they are portrayed in crystalline form, using polarized light.

## The body's very own tempters

Scent is now recognized as a key factor in triggering and deepening a relationship. The pheromones our bodies release are known to influence the behavior of other members of our species. If two people are attracted to each other's scents, they probably form a strong bond, while if their aromas do not combine well, their relationship probably has very little chance of long-term success.

In recent years researchers have found specific receptor genes for the pheromones. They have identified a special little part of the nasal mucous membrane, known as the vomeronasal organ, that is able to register pheromone signals with the aid of receptors. This organ transmits them to the brain, where they are transformed into an impression or a feeling. The production of pheromones is dependent on hormone production in the body. The composition of pheromones in women in their fertile years is different from that of girls prior to puberty and of women after menopause.

Compared with ordinary odors, pheromones are often more subtle, functioning as subconscious stimulants. They penetrate our clothing, our car upholstery, and the fabrics at home and at work. They may exert a subconscious effect on those around us. Thus human beings, like many animals, mark their territory. When we have a cold, we are less able to perceive the pheromones around us. The cold goes away, but a stuffed nose produced by a chronic allergy may lastingly disturb our perception of smells.

**A nose for love**

Our noses help us capture subtle signals. The nasal mucous membranes register everyday scents, and the vomeronasal organ, located on either side of the nasal bone, contains receptors specifically for pheromones. Below we see the vomeronasal organ surrounded by female pheromone molecules magnified hundreds of thousands of times larger than actual size.

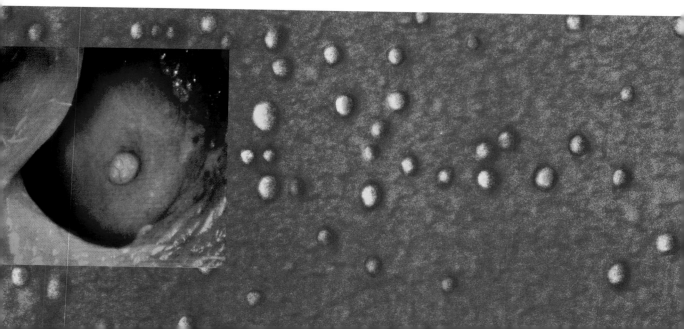

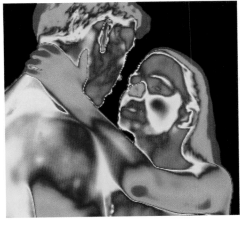

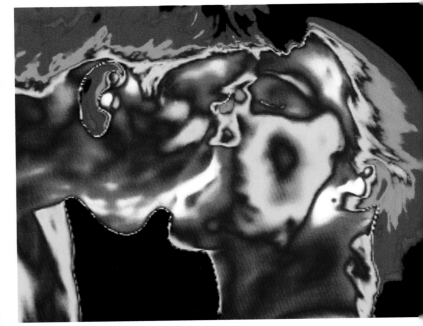

## Life begins with a kiss

Red is not only the color of love, it is also the color that a thermal camera registers when embraces and kisses make the blood flow faster in the bodies of two people who are in love. Light red tones indicate the hottest spots, and blue the coldest.

Some parts of the body, such as the genitals, nipples, earlobes, lips, and throat are more sensitive to touch than others.

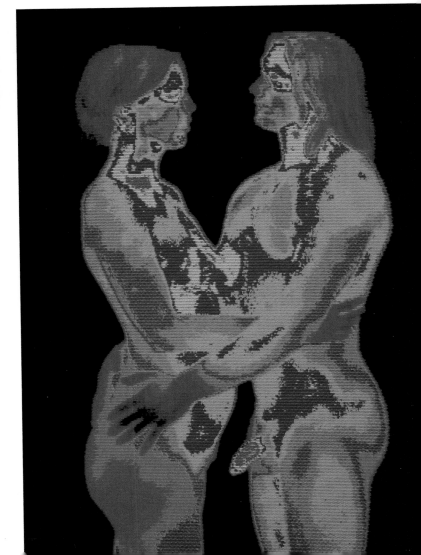

## Similar and yet unique

You are a human being too! How can I tell? You may be tall or short, thin or fat, dark or fair, man or woman, but both you and I belong to the biological species homo sapiens, with a shared genetic code that distinguishes us, for example, from apes, from pigs, and from birds. The great apes, chimpanzees, gorillas, and orangutans, are our closest relations; our genetic codes differ only marginally from theirs. We are also genetically very similar to swine, or the porcine family. The differences from one human being to the next are even smaller, measurable in tenths of a percent, but they are still big enough to make each individual unique. The only people who have precisely the same genetic material are identical twins. Human beings of the same race are genetically very similar; the closest likeness is among members of the same nuclear or extended family, who often share the same hair and eye color, height and weight, and even health status and life expectancy.

In recent years genetics has become a central focus in biology. Although for a time every trait was thought to have a genetic explanation, today we know that genetics and environment are in constant interaction. We also know more about how our environment affects us, and that the environment we experienced at the embryonic and fetal stages of our development has influenced our later lives. We are now very much aware of how important it is for pregnant women to think about what they eat and do.

Our knowledge and awareness of the significance of genetics, both for mankind as a unique species and for each individual, has increased enormously over the latest decade. Scientists have now described the entire human genetic sequence, containing nearly forty thousand genes, each carrying unique information. But we still know relatively little about what these genes do, what they imply for each of us, and how they either cooperate with or oppose one another in our bodies. Still more remains to be learned about how the environment affects the ability of each individual gene to express itself. Not until we learn much more will we have practical use of our knowledge of the genetic code. Only then will we be sure of the potential benefits of this knowledge—and also its possible risks—in relation to human health.

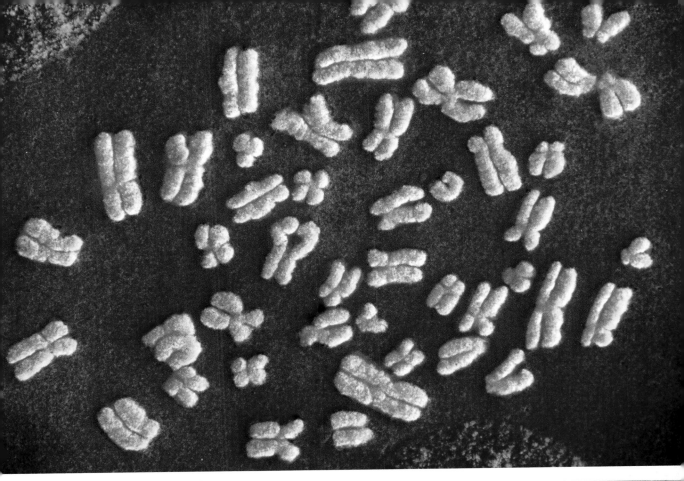

### Our genetic legacy

Our entire genetic code is contained in the chromosomes that exist in every cell in our bodies (top). Inside each chromosome there are specific spaces assigned to more than a thousand genes. Somewhere in this tightly organized system is a gene for each trait, such as the red hair that the baby has inherited. Some traits have to be inherited from both parents in order to be expressed in the child.

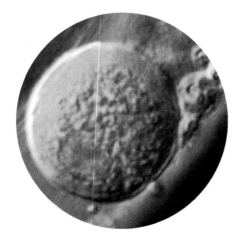

## The human code

Every cell in the human body has a nucleus, which is where our genetic material is stored. Our genes are packed into forty-six chromosomes, arranged with great precision. This structure, with forty-six chromosomes and approximately forty thousand genes, is common to all human beings. But there are small variations within the structure that determine the characteristics of each human being, and these little differences are what make every individual just slightly different from every other, in terms of appearance, talents, behavior, and so on.

Since the genetic material in every cell of each human being looks identical, the details of an individual's genetic composition may be determined by examining any single cell. These techniques are used today to trace hereditary disposition for certain diseases, as well as to trace criminals or to determine a suspect's guilt based on DNA evidence from a crime scene.

Genetic material consists of DNA molecules in the shape of an extended double helix, intertwined spirals of chemical building blocks often designated by the letters A, C, G, and T. The different combinations of these letters give a very large number of different messages. If the DNA chain were laid out flat, it would be 1.8 meters (2.3 yards) long. It is amazing that this chain, containing an astonishing million pieces of encoded information, is somehow enclosed in every nucleus of every cell in our bodies!

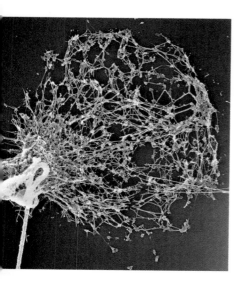

**Long information chain**

Upon close scrutiny any single chromosome displays various bands, some darker and some lighter. Each band (active genes) contains the long spiral DNA chain, stored in a specified place, coded for production of a specific protein substance. At left, the DNA structure of the head of a sperm, showing its complex construction.

18

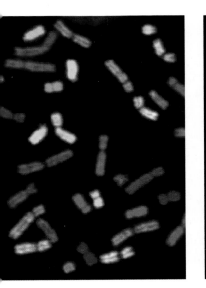

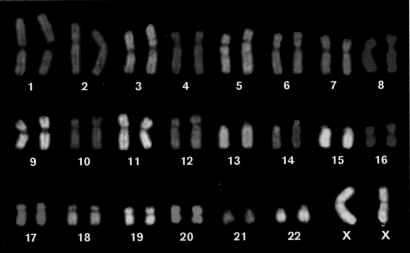

### The basic plan

A woman's twenty-three different types of chromosomes are displayed in different colors (chromosome painting). Only at the very instant when a cell divides is it possible to distinguish individual chromosomes clearly, under a microscope. Each chromosome has two arms, which are connected in the middle by a little rounded structure. Each chromosome also has a specific appearance. Today, using a computer, it is easy to identify and sort them into pairs, but only an expert can discover and assess any aberrations.

Our cells multiply by division, and each time a cell divides, two new ones, with exactly the same genetic material, are created. In every second in time, in every part of our bodies, throughout our entire lives, thousands of new cells, identical to the old ones, are being created. Once the cells of a bodily organ age, they expire, in accordance with a special pattern (known as programmed cell death, or apoptosis), and are replaced by new ones, keeping our bodies young and strong for many years. Ongoing research on the genetic codes that govern the ways in which our bodies age suggests that it may be possible not only to prolong our lives but also to enable women to bear children at increasingly late ages.

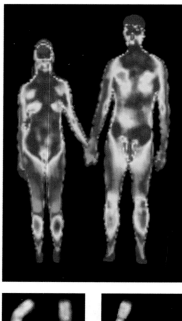

**Y makes all the difference**

Men have one extra chromosome type that women do not. The Y chromosome, which replaces one of the X chromosomes in a male, is the smallest chromosome in the human being.

## X and Y

Sex cells differ from the other cells in the human body in that at the moment of fertilization they contain only twenty-three chromosomes each. When egg and sperm fuse, the same number of chromosomes come from the man and the woman, bringing the total back up to forty-six chromosomes, in twenty-three pairs. The first twenty-two pairs of chromosomes are the same in both sexes. The twenty-third pair is the unique one, consisting either of two X chromosomes in women, or one X and one Y chromosome in men.

Immature eggs contain forty-six chromosomes, like all cells in the body, and the twenty-third pair is always two X chromosomes. But a few hours before ovulation, when the egg is almost ready to be fertilized, the number of chromosomes is reduced by half. Thus twenty-three chromosomes remain in the egg, and twenty-three are rejected into a little polar body located just outside the cytoplasm of the egg (the material surrounding the nucleus) but inside the shell.

Although immature sperm contain forty-six (twenty-three pairs) chromosomes, this number is also halved during the process by which the sperm matures. The twenty-third chromosome pair in a man consists of one X and one Y chromosome. When the sperm splits into two parts, one contains an X chromosome and the other a Y. Thus half the mature sperm contain the genetic traits for girls and the other half for boys.

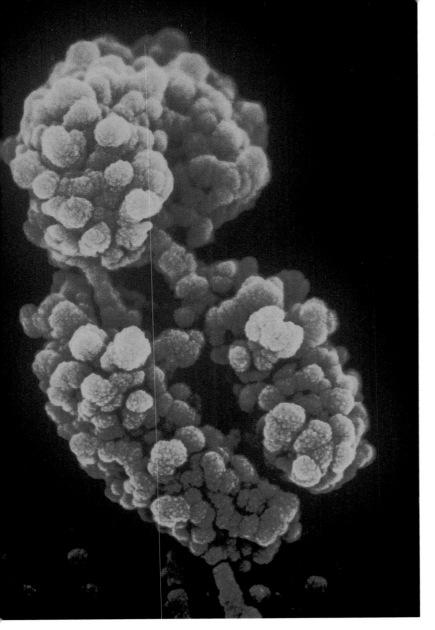

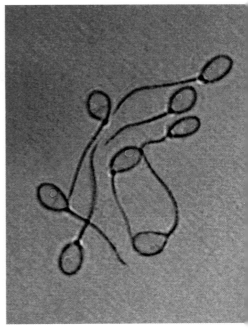

## Two kinds of sperm

Each individual sperm carries either an X or a Y chromosome. If it were possible to separate X from Y sperm with great precision, we could determine the sex of an embryo at the very moment of conception. Many such attempts have been made by more or less serious researchers, but to date no one has successfully separated X from Y sperm. So Mother Nature still holds the trump card in her hand regarding this crucial selection process.

The bright green patch marks a Y chromosome. The X chromosome has a red marker.

## Size is not an issue

The large, plump X chromosome (top left) makes the much smaller Y chromosome appear subordinate. Despite its modest size, however, the Y chromosome plays a very significant role.

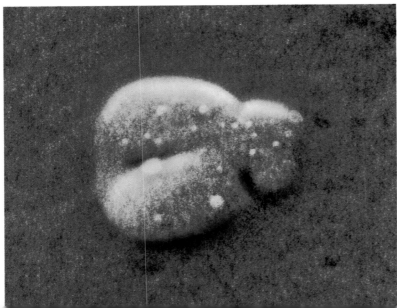

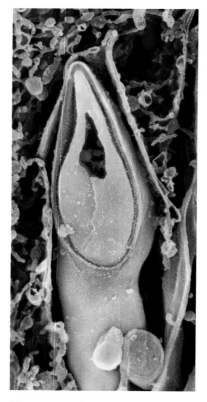

**The sperm—the male sex cell**

Beginning in puberty and continuing until late in life, a man's testicles produce millions of sperm every day. Each sperm contains a set of the man's unique genetic characteristics. The genetic information is tightly packed in the head of the mature sperm (seen here in cross-section).

## The expectant father

In both animals and human beings, the male of the species tends to be physically stronger than the female. Throughout history, too, the role of the head or chief of a group of people, tribe, or family has tended to be a man. In contemporary Western society, however, where physical strength is less of a premium attribute and intellectual capacity, social competence, and knowledge are more highly prized, male dominance has decreased. In reproduction the function of the man, although essential, has always been less significant overall and of shorter duration than that of the woman.

Sperm, or male sex cells, are much smaller than eggs. A sperm consists of a head, which contains the genetic mass, a midsection, and a long, thin tail. The mission of the sperm is to deliver the man's genetic material to the egg, in the process determining the sex of the unborn human being.

Primitive sperm cells, known as spermatogonia, are already in place in the testicles of a newborn baby boy. Hormones from the pituitary, a little gland below the brain, govern the sexual development of boys and trigger production of viable sperm. Why do boys become sexually mature—that is, begin to produce sperm capable of fertilizing an egg—at the age of about twelve or thirteen? There is not yet a clear answer to that question, but we do know that before that time the thymus gland impedes the sexual maturation process, which is then initiated by a complex interaction of many factors, including adequate nutrition supply and various growth hormones and hormones from the adrenal cortex. Genetic factors also play an important role. Luteinizing hormone (LH) helps the testicles to produce testosterone, the male sex hormone, and follicle stimulating hormone (FSH) triggers both sperm production and the maturation of the sperm, making them viable for fertilization. Hormones, particularly testosterone, govern boys' bodily growth, muscle structure, development of external sexual organs, voice change, whiskers, and development of body hair.

The interaction among the hormones is delicately balanced; disturbances may result in permanent damage. This explains why the use of testosteronelike substances, such as anabolic steroids, by young athletes may have dire consequences, including fertility problems later in life.

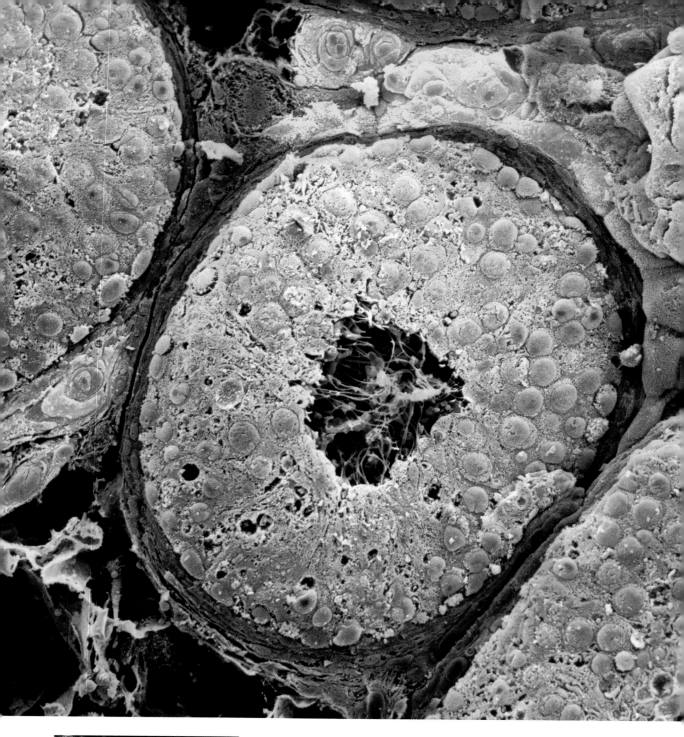

## A sperm and hormone production facility

Male sex cells develop in the seminal canals of the testicles, under the influence of hormones from the pituitary gland. Each canal has a central channel in which the newly produced sperm lie waiting to be transported to the epididymis. The testicles also produce the hormone testosterone (left), in what are known as the Leydig cells (in the large photograph, the little groups of yellow cells between the canals) aided by a hormone secreted by the pituitary gland.

## One thousand sperm per second

While a woman generally develops only one egg per month, a man produces billions of new sperm every month. Every time he ejaculates, his body produces 2–6 milliliters (0.3–1.0 teaspoons) of seminal fluid, containing up to five hundred million sperm. This production continues in some men until the age of eighty, although the speed and quality of sperm production gradually declines. The sperm production process also appears to be considerably less efficient than women's egg production. Many sperm lack either adequate propulsion ability, endurance of movement, or—worst of all—correct genetic information. Fully 85 percent of the sperm a man produces may be defective in some way, without negative effects on his fertility. From a genetic viewpoint, sperm contain far more aberrations than do eggs. So the egg maintains the unique human code, while the sperm are constantly introducing variation. In other words, sperm are crucial to human evolution.

The body takes over seventy days to make a sperm. They are created in the testicles in a meandering tangle of seminal canals several hundred meters long. Along the walls of these canals the sperm begin to grow, under hormonal control. In the small cell clusters surrounding the seminal canals, testosterone, the essential male sex hormone, is produced.

Scientists in several countries have recently noted an alarming decrease in human males' sperm count. Intensive research into the reasons is ongoing; they may include air pollution and environmental toxins in the food chain. Stress and toxins in certain work situations are also considered possible causes. The treatment of livestock with antibiotics and hormones that then remain in the meat we eat may also be a factor.

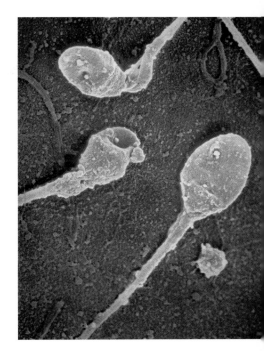

**A swimming cell packet**

Sperm, with their heads, tails, and propulsion abilities, are among the most remarkable types of cells. They are produced in mind-boggling quantities and at great speed. But only about one mature sperm in ten is completely perfect and thus capable of fertilizing an egg.

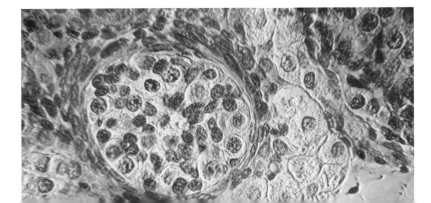

**Primitive sperm**

Sperm begin to develop even when the male is still an embryo, but not until a young man becomes sexually mature does he produce sperm that can fertilize an egg. In this cross-section of the testicle of a fetus in week 18, the seminal canal is visible in the middle.

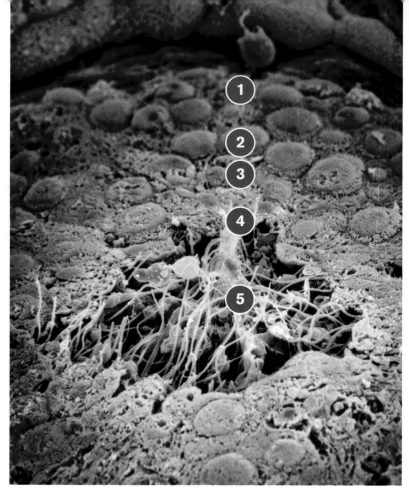

## A tangle in the testicles

Coils of seminal canals fill each testicle to the brim. In a healthy man, nearly a hundred million sperm are produced in the canals every day. The photos to the right and on the next page show the various phases of the development of sperm toward maturity.

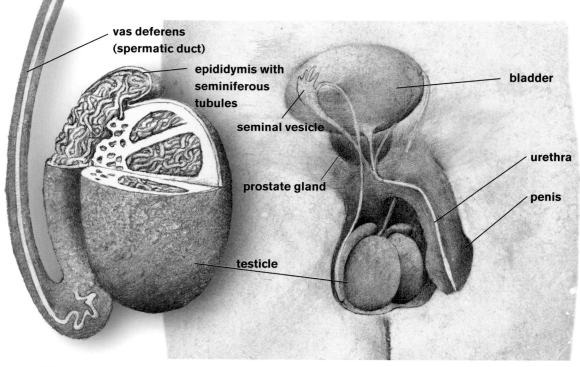

vas deferens
(spermatic duct)

epididymis with
seminiferous
tubules

seminal vesicle

bladder

prostate gland

urethra

penis

testicle

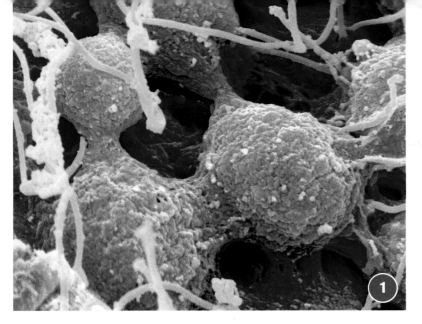

**The immature sperm divide**

An immature sperm contains forty-six chromosomes. Early in the production process it divides itself into two spermatids, which have twenty-three chromosomes each so as to be able to unite with the twenty-three chromosomes of an egg. In the photo to the left, a bridge remains between two new spermatids.

## The sperm maturation process

Sperm production occurs in the walls of the seminiferous tubules, where immature sperm are produced. As the sperm mature, they move into the channel of the seminiferous tubules tail first, head last.

At this point the sperm are still unable to propel themselves, but are pulled passively along in the secretion that flows through the canals. These canals meet, merging into wider channels, and eventually the sperm reach the central sperm holding tank known as the epididymis, where the final maturation process takes place, where the tail begins to function and the sperm become mobile.

Sperm do not survive forever. If they are not ejaculated, they eventually die, making way for newly produced sperm.

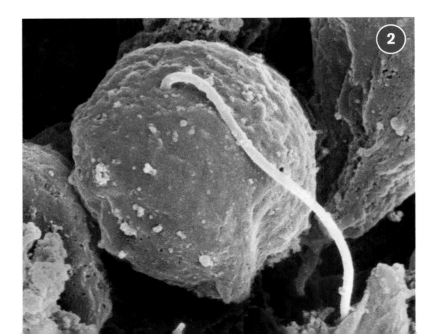

**The tail grows out**

Here a spermatid is shifting toward the middle of the seminal canal, and a tail is now developing on this relatively large, slightly cumbersome cell.

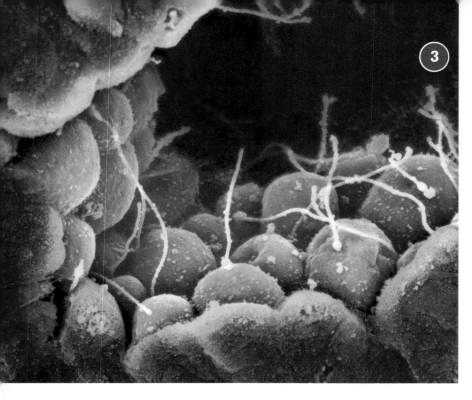

### Deep within the system of passageways

Near the center of the seminal canal multitudes of semimature sperm wait to reach maturity. Most of them have tails.

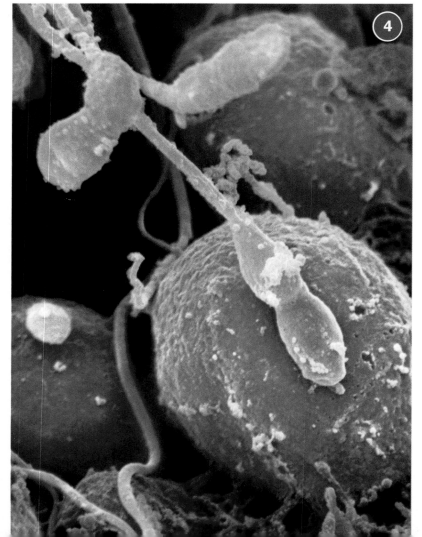

### Functional design

As a sperm approaches maturity, its appearance changes. The rounded cell becomes more streamlined, a shape better suited to the task at hand: swimming competitively through a woman's reproductive system.

### Ready for delivery

Still in the seminal canal are hundreds of sperm with tails that have grown ever longer; there appears to be an imminent risk that they might become tangled up. A slow steady flow of fluid transports the sperm to the catchment area in the epididymis, where the tails finally become capable of propulsion.

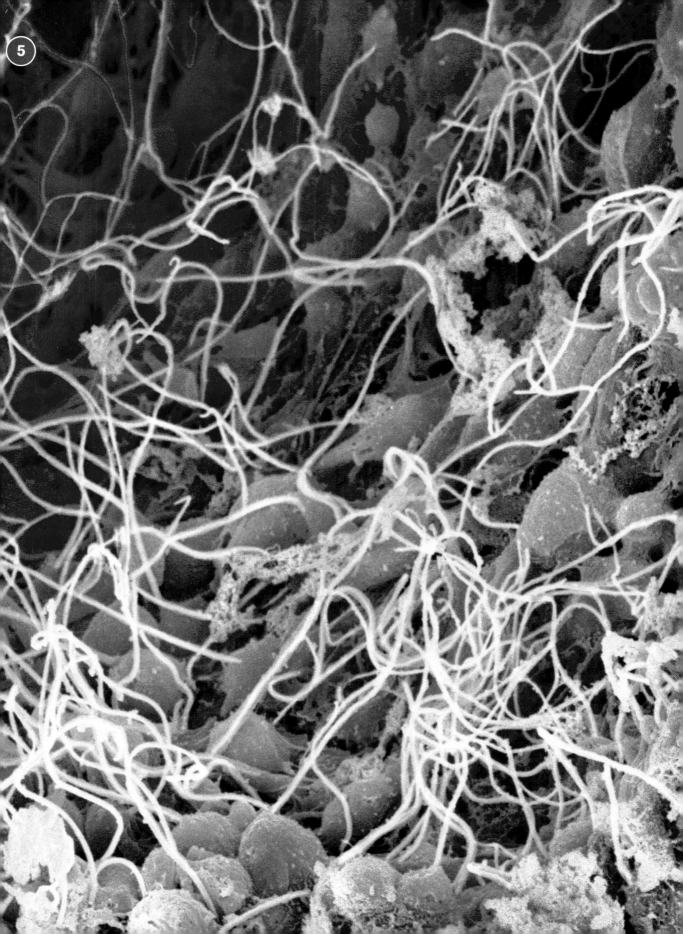

**The egg–the woman's sex cell**

The woman's egg (oocyte) is the equivalent of the man's sperm. It contains a unique genetic code but, unlike sperm, lacks billions of competitors. Each month in a woman of fertile age, a single egg in one of her ovaries prepares to be fertilized.

# The expectant mother

After both the man and the woman have passed their own genetic material on to the new individual, the woman plays the main part until birth. In her body the first cell divisions take place, and a new little being is formed. Her reproductive system, which prepares for fertilization and pregnancy every month for about thirty-five years, is optimally adapted to nurturing an embryo and carrying it to term.

All the eggs the woman will ever produce have been in her ovaries in an immature form since she herself was an embryo, but they mature much later and only one at a time. Before puberty her hormones undergo a change that makes her ovaries begin to produce more estrogen, and the eggs begin to mature. The shape of her girlish body also begins to change, and her entrance into adulthood is confirmed when she begins to menstruate. A century ago young women usually had their first period around the age of fifteen, but today in well-to-do countries such as Sweden and the United States and Britain menarche is much earlier (the average age is 12.5). One explanation is that the timing is governed more by weight than by age. We eat differently today, and girls tend to achieve the critical weight for menstruation (46–47 kilograms/ 101–103 pounds) far earlier than in the past. Of course, there are wide individual and ethnic variations.

When a young woman's first, often irregular periods become more punctual on a monthly basis, she has probably begun to ovulate, and thus can become pregnant. There is therefore quite a gap between the time a woman can reproduce biologically, and the time when, from a social and cultural point of view, she is considered able to become a mother. On this subject perspectives between the developed and developing countries differ. In the former the pendulum has swung so far that many women are over thirty or thirty-five or even older before having their first child.

There are, of course, both advantages and disadvantages to having children later in life. While maturity, experience, and job security may help a woman approach the responsibilities of motherhood with more confidence, a somewhat older woman will likely have fewer chances of becoming pregnant or having the number of children she wishes to have. The risk of miscarriage increases as well, since the genetic quality of eggs deteriorates over time.

## The female hormone

Estrogen, which is secreted by the woman's ovaries, is the most important female hormone (right). It contributes to the development of her figure, the size of her breasts, the softness of her skin, and the thickness of her pubic hair; it also affects some regions of her brain. Estrogen is transported by the circulatory system to different parts of her body. But without special receptors that react to hormonal signals, it would have no effect. There are two different estrogen receptors: alpha receptors (above) and beta receptors (below). These determine the many different ways in which tissues in different organs react to the estrogen.

## *Hormones control the menstrual cycle*

Every menstrual period in a woman's life begins a new cycle, lasting approximately four weeks. The first hormonal impulses come from the pituitary gland. Which hormones are to be excreted, and in what quantities, is determined by centers in the lower part of the brain that are in direct contact with the pituitary gland.

The brain has a great influence on the menstrual cycle. Many women, when they are under severe stress or are very worried, experience periods that are delayed or that even stop altogether for one or more months. Anorexia, too, can result in loss of menstruation. Even a small change in a woman's life, such as dieting or vacationing, can result in the temporary disappearance of ovulation and menstruation. So can a very strong desire to become pregnant. Young athletes and dancers are especially vulnerable. The stress that their physical activity imposes on their brains and bodies can affect their biological clock, which has to be operating properly in order for a woman to become pregnant and carry a baby. But if she is in reasonably good psychological balance, her brain will instruct the pituitary to excrete hormones, which will be carried to her ovaries by the circulatory system. The ovaries react to this stimulus by increasing production of estrogen.

Normal menstruation usually lasts three to five days, during which time the woman loses about 40 milliliters (1.4 fluid ounces) of blood, although most women would probably guess that they lose more. The woman bleeds because the layer of mucous membrane lining the uterus, into which a fertilized egg could have burrowed, is being rejected and replaced with fresh cells for the next cycle. It takes about a week for the membrane to be rebuilt, to achieve the proper thickness, and to develop its fine network of tiny blood vessels.

At the same time, hormones from the pituitary have signaled to the ovaries to make a few immature eggs begin to grow and mature. Usually a few of all the possible little clusters react fastest to these signals, and approximately four or five eggs start to mature. Usually only one matures (two for fraternal twins) sufficiently to ovulate. Whether ovulation takes place in one or the other ovary appears to be a random matter. Often the ovaries do not take turns producing eggs with complete regularity. Should a woman need to have one ovary removed, the remaining one will then ovulate every month.

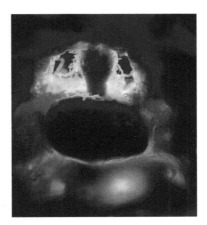

**The conductor**

The pituitary gland is a small hormone-producing gland linked to the lower part of the brain via nerve fibers. It secretes a number of hormones that govern growth, thyroid function, and the production of eggs ready to be fertilized. In women disturbed function of the pituitary may cause ovulation to fail. In men the pituitary is important to sperm production.

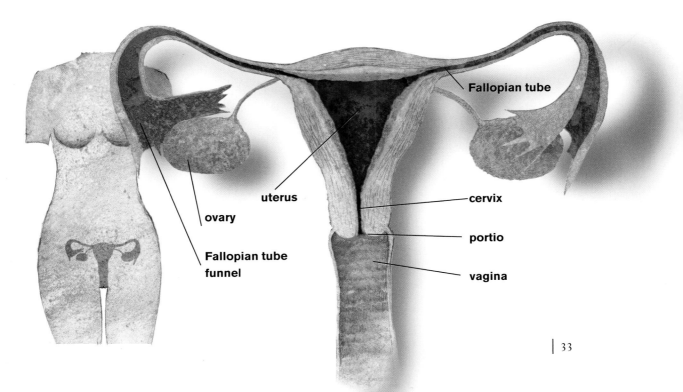

ovary

uterus

Fallopian tube funnel

Fallopian tube

cervix

portio

vagina

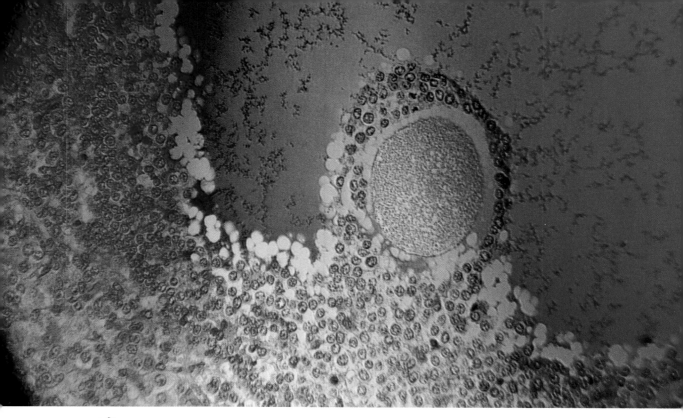

## The body prepares

As ovulation approaches, the woman often notices that the amount of mucus discharged from her vagina increases. This mucus comes from the cervix and increases at ovulation, when it becomes crystal clear and stringy. Only when the mucus has this particular quality will the sperm be able to pass up through the cervix. Close daily examination of cervical secretion is one method that women in many parts of the world use for "natural family planning." This technique may be used both as a form of birth control (in which case the couple avoids having intercourse for a few days every month) and as a good way to know when the woman is ovulating (for a couple wishing to become pregnant).

There are other signs of ovulation as well. Some women always feel it in their backs, while others may have a day or two of spot bleeding. The woman's body temperature rises by about half a degree. She can check her temperature in the mornings as a way of determining when there is a chance of becoming pregnant.

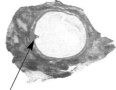

**The maturing egg**

A few days before ovulation a follicle in one of the ovaries begins to develop rapidly, and the oocyte inside moves toward the edge. Slowly the follicle rises toward the surface of the ovary. The large photograph above shows the egg in its follicle just before ovulation.

**Might I get pregnant today?**

The closer intercourse occurs to the moment of ovulation, the greater the woman's chance of becoming pregnant. The probability is highest—26 percent—on the day after ovulation.

**Day 0 = ovulation**

| -3 | -2 | -1 | 0 | 1 | 2 | 3 | 4 | 5 |
|----|----|----|----|----|----|----|----|----|
| 0% | 11% | 15% | 20% | 26% | 15% | 9% | 5% | 0% |

## The eggs are there from birth

In contrast to a man, who produces sex cells throughout his life, a woman's stock of eggs develops before she is born, then decreases gradually. In the fourth month of embryonic development the ovaries of the female fetus have already produced the six to seven million eggs that comprise her lifetime production of eggs.

Even before she is born, millions of the eggs expire; this programmed cell death in the ovaries continues steadily after birth. By puberty, when the young woman ovulates for the first time, only about two million immature eggs remain, and by the time she reaches menopause, in her fifties, virtually her entire stock has been depleted. During the entire period in which a woman is fertile and ovulating, her ovaries actually use up only two hundred to four hundred eggs, of which at most only a few will be fertilized and become a child. So a woman has an enormous reserve capacity, even if her surplus of sex cells is less than a man's.

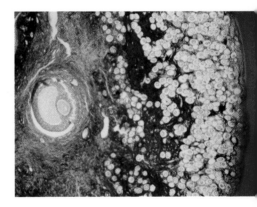

**Immature egg cells**

As a fetus a female has large numbers of immature eggs in her ovaries. This photo shows a fetal ovary during week 30 of pregnancy, with many minuscule egg cells and one larger one, containing a perfectly visible egg.

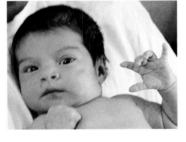

**The ovary of a newborn girl**

**... of a fertile woman**

**... and of a fifty-year-old woman**

**A nonrenewable repository**

A newborn baby girl has a few nearly mature oocytes in each ovary, but they are not viable for fertilization. Throughout a woman's fertile period (some thirty-five years) a maximum of four hundred mature eggs will be released; this figure is often lower since pregnancies, nursing and birth control pills all prevent a woman from ovulating for certain periods. A woman in her fifties will have no follicles left, although her ovaries may go on producing hormones for several years.

## The magnificent egg cell

The mature egg, viable for fertilization, is one tenth of a millimeter (0.0039 inch) in diameter, almost large enough to be visible to the naked eye. The ovum is the largest cell in the human body, a giant in the cellular microworld, but its nucleus is far, far smaller—in fact, no larger than the head of a sperm, or about the same size as other cells in the human body. The nucleus is where the woman's genetic material is stored. All around the mature egg there is a large number of cells, necessary to provide it with nutrition and oxygen for the first few days if it is fertilized in the Fallopian tube.

Everywhere in the cytoplasm of the egg are mitochondria, little fuel packs that supply the egg with energy and other necessities. There are about a thousand of these mitochondria, which also contain genetic material in the form of special DNA molecules. If this DNA happens to be abnormal, congenital illness may appear in the mother and may be transferred to the baby. As a woman ages, the workings of the mitochondria in the eggs deteriorate. This is considered the main reason older women have more difficulty becoming pregnant and miscarry more frequently. The other non-sex cells in our bodies also contain mitochondria that are involved in our energy production and aging processes.

**Ripe for fertilization**

At the time of ovulation the egg rids itself of half its genetic information, sending it off to a peripheral polar body. Twenty-three chromosomes are left in the kernel. The egg is surrounded by a layer of nutrient cells that will be transported along with it after ovulation into the Fallopian tube. Now the egg is ready to meet the sperm, if the opportunity presents itself.

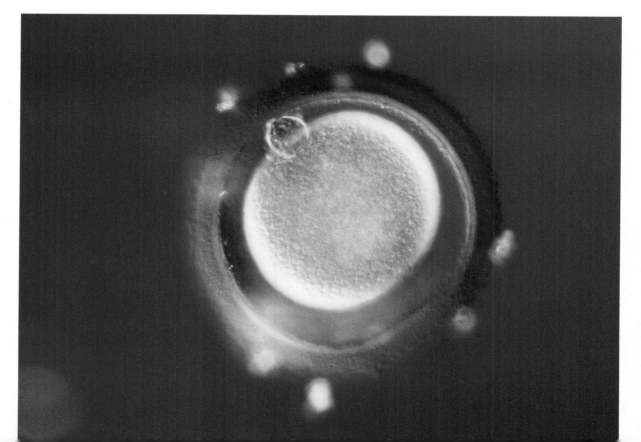

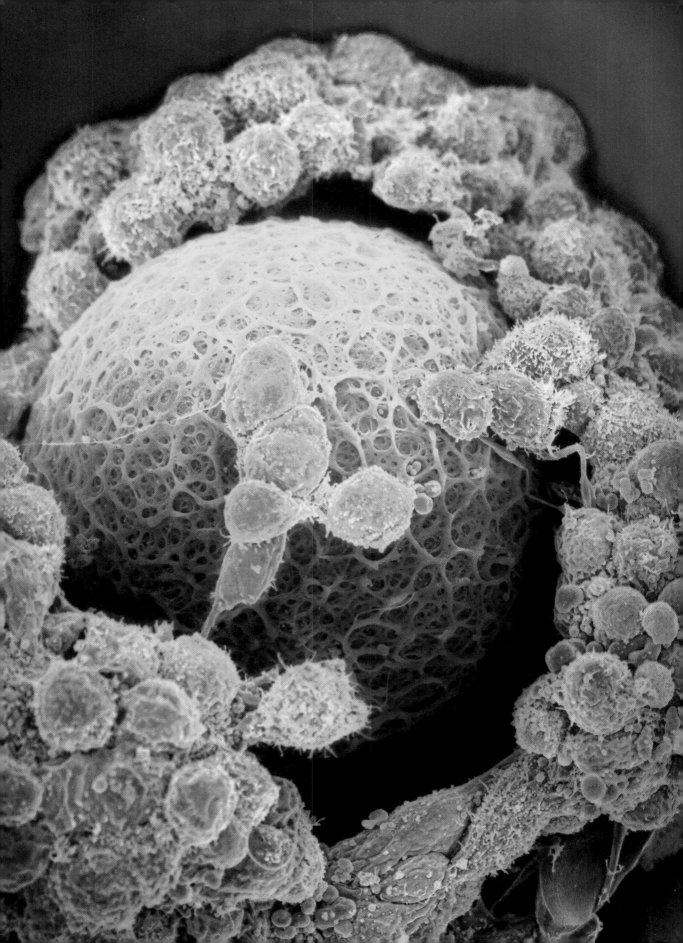

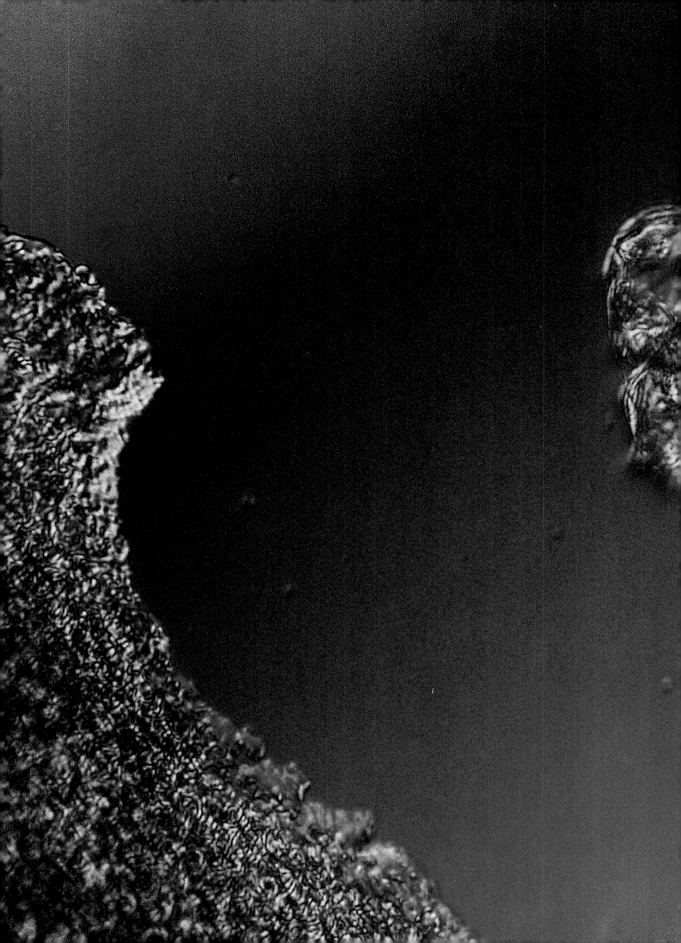

# Fertilization and Conception

For new life to be created, a complex jigsaw puzzle of events must fall into place, each piece fitting perfectly. Few processes in human history have been so shrouded in myth and superstition as conception. Only in recent years have researchers been able to show in detail how it happens.

**The egg is ejected**

Although it looks like a volcanic eruption, the process of ejecting the egg from the follicle is really quite slow. One follicle in the ovary—which has hitherto enveloped the egg in its protective chrysalis—suddenly ruptures, and the egg is pushed out, in a cloud of nutrient cells. Just outside, the sensitive fingers (below) in the funnel of the Fallopian tube are waiting, open toward the ovary, to catch the priceless egg.

## Ovulation

A woman normally ovulates once a month, about two weeks after her last period. This is the only time when intercourse can result in fertilization. In other words, sperm have the opportunity to accomplish their mission for only a few days every month. Fertilization takes place in the Fallopian tube, but it makes no difference whether the sperm or the egg gets there first. When the woman ovulates, sperm may be lying in wait in the folds of the mucous membrane lining the tube, where they can survive for several days. An egg is more sensitive: once it is in the tube, the sperm have to appear—and fertilization must take place—within forty-eight hours if there is to be a viable chance of pregnancy.

The ovulation process begins when a small number of ovarian follicles begin to grow. Usually one develops fastest. By about two weeks after the beginning of menstruation, that follicle is approximately 2 centimeters (1 inch) across or somewhat larger, and it contains an oocyte ready to be fertilized. The follicle ruptures, and the liquid that fills it (some 10–15 milliliters/0.6–1 tablespoon) is released, along with millions of cells that have been secreting estrogen. The oocyte itself, the precious cargo of the follicle, is in this swarm, buffered by thousands of protective cells responsible for its immediate surroundings and its nutrient supply.

When the oocyte has been released from the follicle, instead of tumbling down into the abdominal cavity, it is almost always caught up in the supple, ingeniously constructed outer funnel of the Fallopian tube, also known as the uterine tube or oviduct. If a woman has scar tissue from an infection on this part of her body, this sophisticated mechanism is easily disturbed.

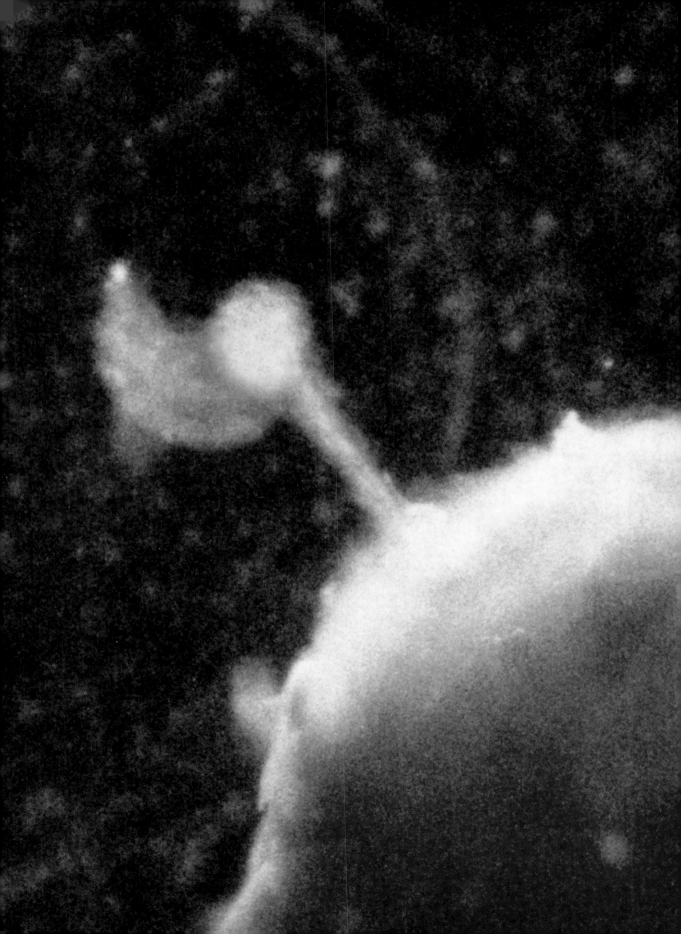

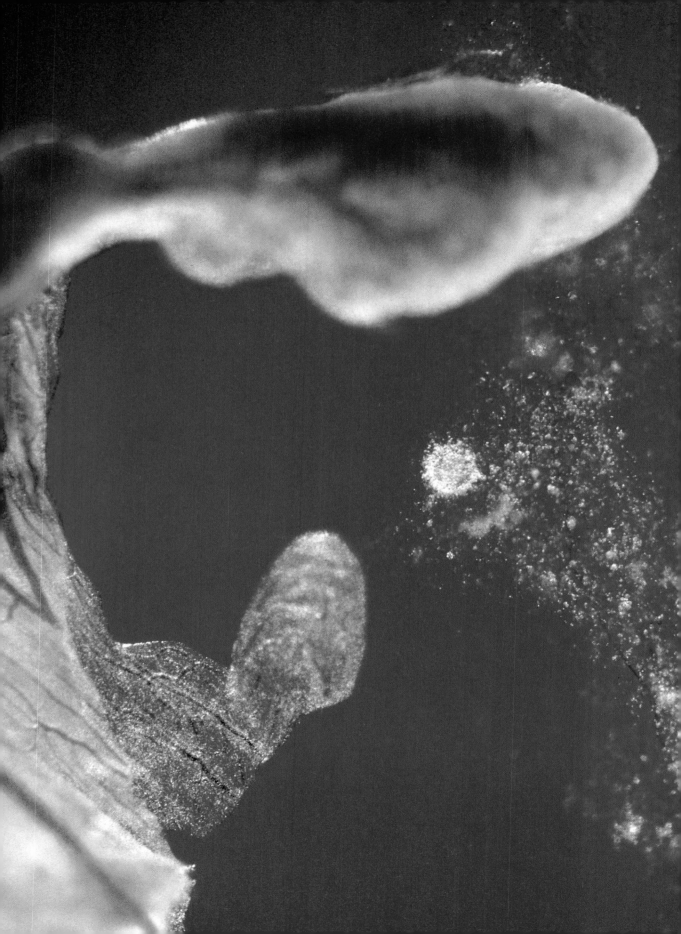

## *The egg is caught*

From the opening of the Fallopian tube funnel, the egg is transported into the sheltered environment of the tube, or oviduct, itself. There it is prevented from dropping into the abdominal cavity by the beating motions of the minuscule cilia, or hairlike projections, in the mucous membrane. The egg remains in this open, wider part of the oviduct in anticipation of possible fertilization for about forty-eight hours. It does not always lie still but rolls slowly along the membranes, awaiting its male partner. This snug area in the oviduct provides optimal protection for both egg and sperm. The egg comes to full maturity there. If no sperm arrives, or if the egg's encounter with the sperm fails to result in an embryo, it will simply continue down through the oviduct into the uterus and then out past the cervix, to be discharged through the vagina. Then in about ten days, the woman will menstruate.

Meanwhile the mature follicle, once it has ruptured, goes on to play another important role in the ovary as the corpus luteum. Having secreted large amounts of estrogen, it begins manufacturing a hormone that is chemically related but has very different properties. This is progesterone, and it goes out into the bloodstream and influences the outer cell layer lining the uterus, preparing it to receive a fertilized egg.

**A new mission for the follicle**

The swollen follicle has collapsed; a hole gapes in its middle from which the oocyte was thrust forth. But the follicle's job is not finished. It is now transformed into a corpus luteum, and its cells begin producing progesterone.

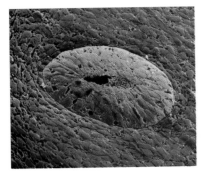

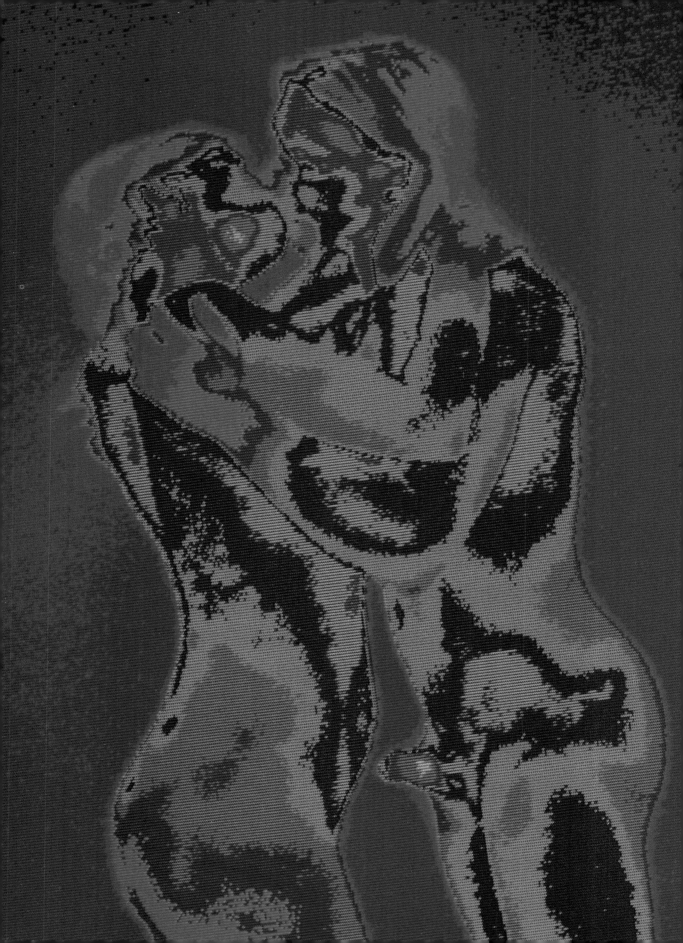

## Foreplay

The egg rolls around in the Fallopian tube, bouncing softly off the thousands of tiny cilia that move it gently along its course. The surrounding layer of nutrient cells begins to dissolve, owing both to friction and to the effect of enzymes in the tube's secretions. The egg is thus more lightly dressed in anticipation of approaching sperm.

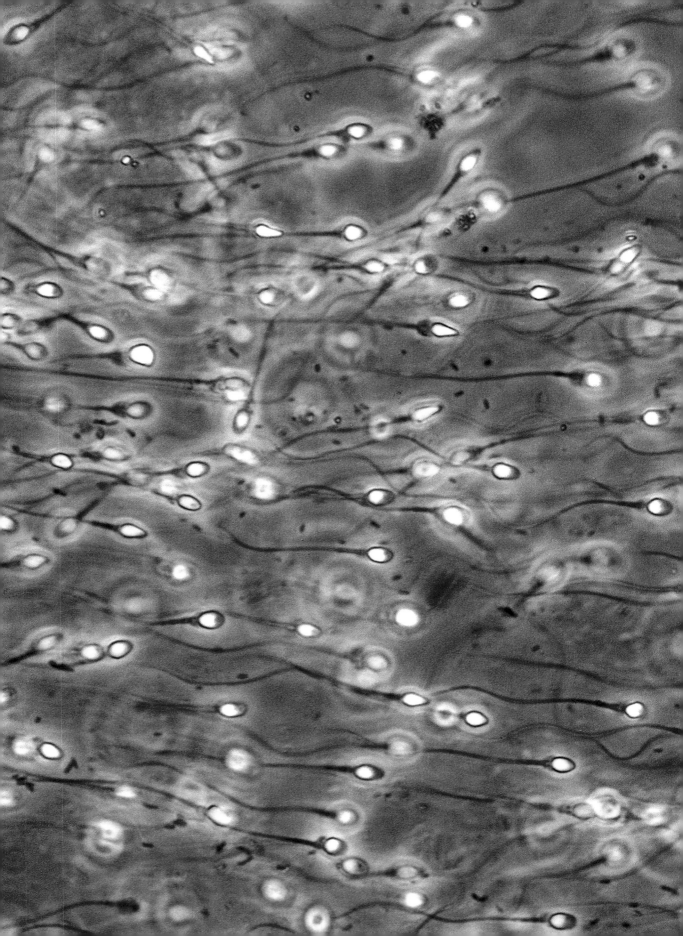

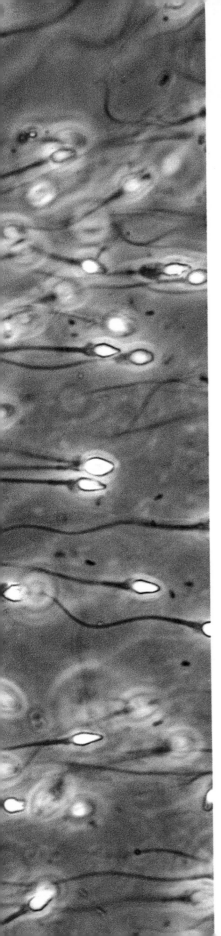

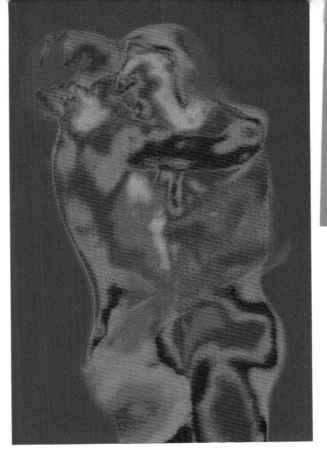

## *The race*

During intercourse, when the man's pelvic muscles contract in orgasm, his sperm shoot out of the epididymis into the urethra, mixing with the secretion from the prostate gland. This secretion contains substances that enable the sperm to move more effectively toward the woman's egg. His seminal fluid shoots out into the vagina: the sperm and secretion from the prostate first, followed by a gelatinous fluid that flows from the seminal vesicles to the urethra.

In an erection, the blood flow through the penis increases, and the erectile tissue (the corpora cavernosa) fills with blood and expands. After ejaculation, the blood flow subsides, and the tissue returns to normal size.

### They're off!

After ejaculation a swarm of sperm enters the vagina and takes off for the cervix. This is the initial heat in the race for life, with up to five hundred million competitors. Who will win? It is not enough to be fastest and strongest; one must also find the simplest, shortest route.

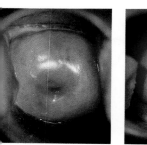
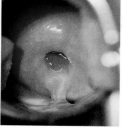
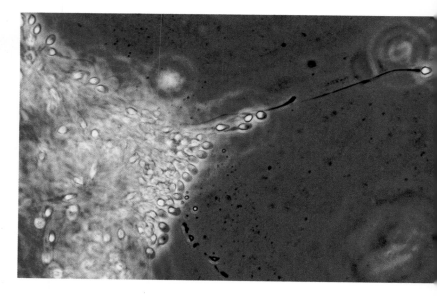

**Portal to life**

The cervix, protruding down into
the top of the vagina, is the eye of
the needle that the sperm have to
get through on their way toward
the egg. At the time of ovulation
the vagina produces a large amount
of a crystal-clear discharge, the
function of which is to screen the
sperm and perform the first selection process. Sperm that are too
weak become lodged in this mucus,
unable to pass through. The portal
to the uterus is open only a very
few days out of every month, after
which it is plugged shut again by
thick mucus that shuts out not only
latecoming sperm but also all kinds
of bacteria.

## Swimming the channels toward the uterus

Some of the sperm that enter the vagina flow right back out again,
while others rest on the mucus, and in the many folds of the mucous membranes. Their tails propel them as, swimming for all they
are worth, they drive deep into the channels in the mucus. These
passages are often narrow and congested. During ovulation and for
the following few days, the woman's estrogen helps unblock these
channels, opening them wide. The sperm course toward the opening of the cervix. Their window of opportunity is short, because
the cervix is open only briefly. After just a day or so, this mouth of
the uterus closes up, and a thick plug of mucus impedes the eagerly
swimming sperm from further progress.

**A narrow passage**

Once the sperm have survived
the first manly test—the mucus in
the vagina—they have to make their
way up one of the narrow channels
that form in the saccharine secretion of the cervix.

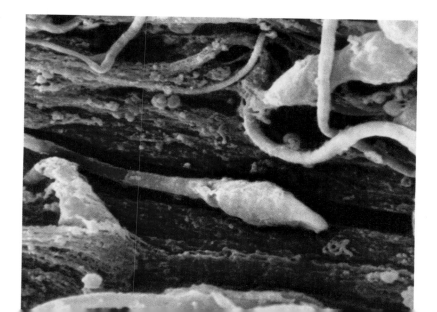

## The engine and fuel of the sperm

To accomplish this demanding journey, the sperm require energy. Just behind the head of each sperm is a midsection containing a packet of energy in the form of mitochondria (top). As the sperm swim, the mitochondria consume the sugary substances around them, which provides extra energy. The wriggly tail that propels the sperm forward is a system of microtubules, a brilliant construction, in a cross-section not unlike an electric cable (center). Without these tubules the sperm would not be able to swim.

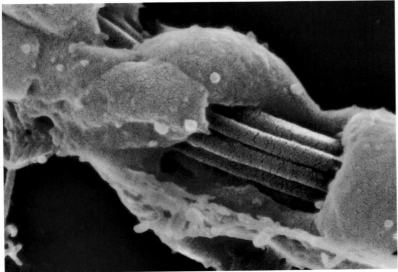

### End in sight

Although fast-swimming sperm may reach the egg in half an hour, others can take days. During the journey the sperm change successively, under the constant influence of substances from the woman's reproductive system. At one point they are "capacitated," after which they are capable of fertilizing the egg. The sperm leading this race is just visible at the very bottom left of the photo opposite.

### Deterrents

Making headway through the rough landscape of the Fallopian tube, which can be compared to a coral reef, is no easy task. Lots of sperm use up all their energy in a futile struggle with the dense mass of cilia and flounder like fish caught in a net. The cilia beat "against the flow," making the journey for the sperm even more of an uphill struggle.

## Upstream through the Fallopian tube

The sperm have a long, laborious journey toward the egg, and as they travel through the cervix and the uterus and up into the uterine tube, most succumb. Hundreds do achieve their goal, although they need to stop and rest along the way. It is thought that some sperm, particularly those from young, healthy men, can remain alive and potent for up to four to five days after intercourse.

Since ovulation alternates, more or less, between the two ovaries, only one Fallopian tube will contain the egg. This one is often a little more open and welcoming, but sperm swim into both Fallopian tubes as they career blindly ahead. Thus some sperm "guess wrong" and end up in the tube where there is no egg that month.

It usually takes several hours for a sperm to make its way through the vagina and into the tube, a journey of 15–18 centimeters (6–7 inches). But under favorable conditions some quick swimmers are able to reach the uterine tube within half an hour. Researchers speculate that sperm transport may be facilitated by the female orgasm. It has been estimated that a thousand swishes of the tail propel a sperm approximately one centimeter forward, and that the energy housed in the midsection of a sperm is sufficient to allow it to swim for hours. It can also gather extra fuel from the environment. Still, it is not certain that any of these fastest-swimming sperm will be the one to fertilize the egg.

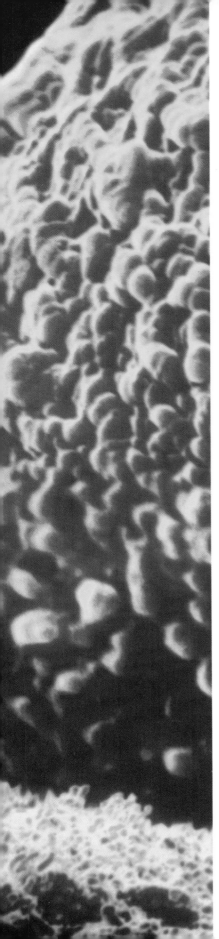

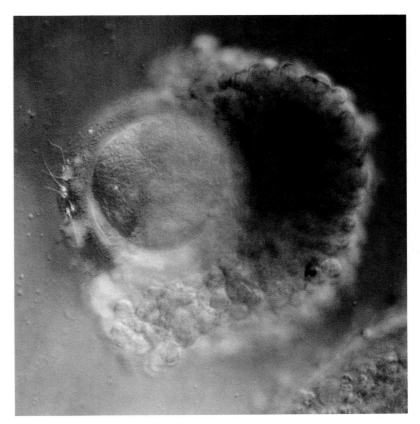

### First encounter

The enormous egg towers before the leading sperm (left), which, after the long swim, has finally reached its goal, along with about a hundred competitors. The egg is still enclosed in nutrient cells; a number of the sperm immediately begin to work their way past these cells and through the outer shell of the egg (above).

*The moment of truth. Will there be enough energy to force through the eggshell?*

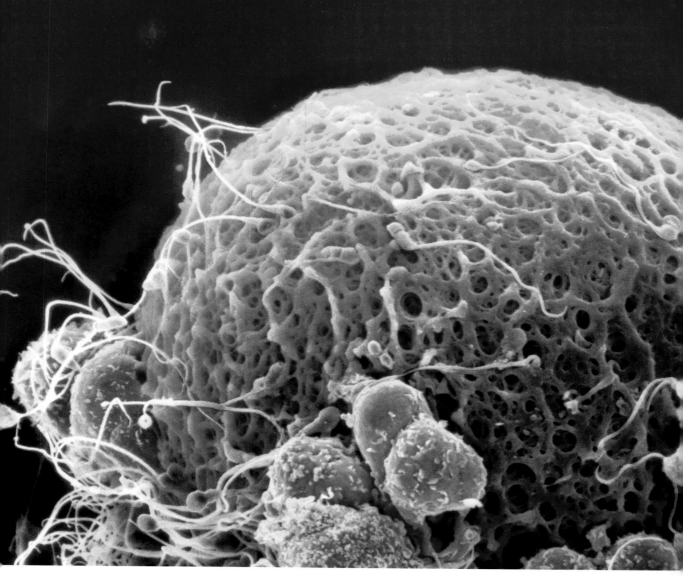

### The final disrobing

The eggshell has many entryways, tiny holes through which sperm can enter (above). First, however, the sperm have to see to it that some of the nutrient cells are removed, making the surface of the egg as nude and as accessible as possible. Although some sperm do not survive this process, they help pave the way for the winner.

## Storming the eggshell

When the quickest sperm have reached the egg, the actual fertilization process begins. The sperm swarm around the large sphere, beginning their frantic efforts to force their way inside. At any given instant there may be no more than a few hundred sperm surrounding the egg, and many of them are impeded by the protective nutrient cells.

The interior of the egg is shielded by a strong shell that is difficult to penetrate and consists not of cells but of a tough, almost hard cohesive material. The sperm will have to penetrate this shell to enter the cytoplasm of the egg, the nucleus where the woman's genetic information is stored. Compared with the huge egg cell,

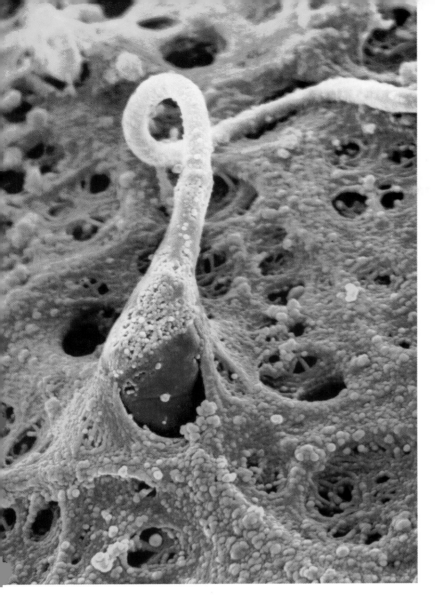

### Passing through the shell

Because of the many craters in the shell, a number of sperm can enter it at the same time (left). The leading sperm may now be only minutes from victory. As it passes through the shell, it sheds its acrosome, an organelle that has covered and protected its head (in red, in the photo opposite). A chemical reaction makes the acrosome vanish, after which the sperm is able to penetrate the egg's interior.

the sperm is very small, but the genetic material stored in its head is equal in size and importance to that in the egg. Each sperm functions rather like a drill: the movement of the tail propels the head around and around like the bit of the drill. It doesn't take long for the sperm to find little nooks and crannies that facilitate its way in. Still, in relation to the sperm, the eggshell is so thick and so hard that it is quite amazing for the drilling efforts of the tiny sperm to make any headway at all.

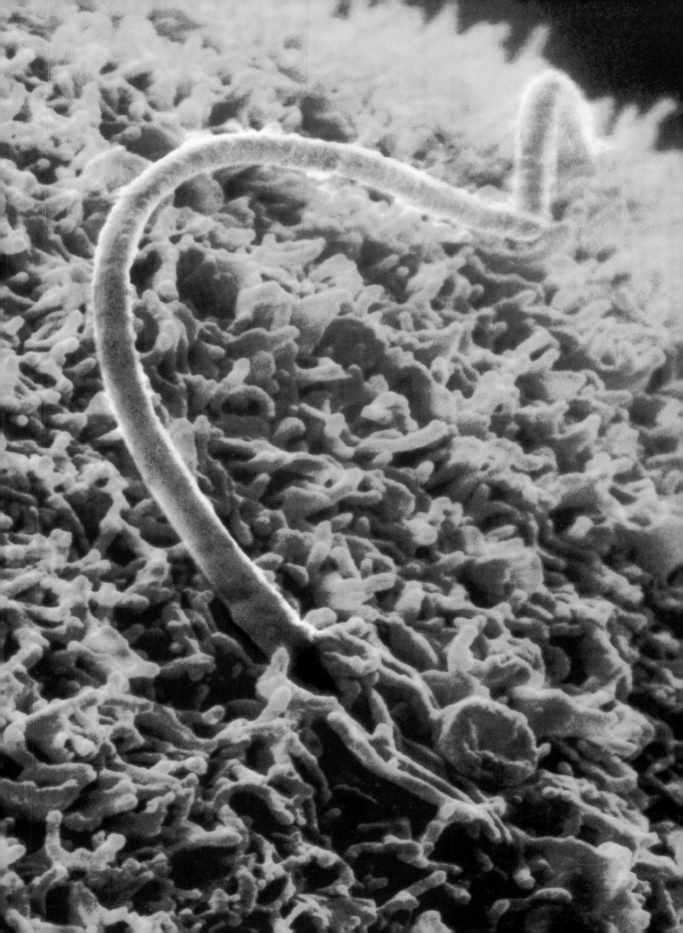

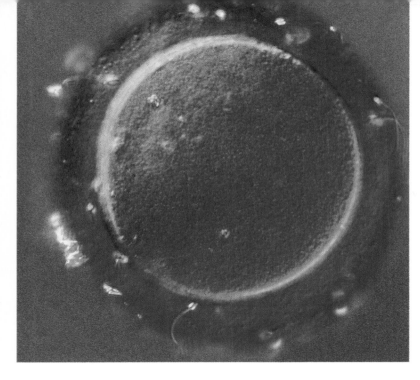

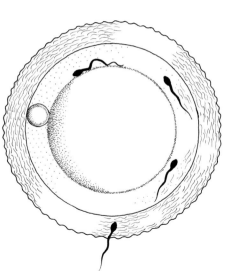

## The final spurt

Several sperm reach the narrow slot between the outer shell and the membrane of the oocyte (above left)—but only one of them can fertilize the egg. The winning sperm makes contact with the "downy" surface of the membrane (opposite), and is rapidly surrounded by the tiny transfer arms covering the surface of the membrane (below left). The sperm seems drawn into the egg, as its head and midsection vanish into the interior (below).

## *The egg lets the sperm in*

After three or four minutes the elite sperm that have taken the lead succeed in penetrating the outer shell and swim on into a fluid-filled space within the egg itself. Suddenly one of the sperm comes to a halt; its head latches on to the membrane surrounding the egg cell. Within a very few minutes it has penetrated the egg: the victor!

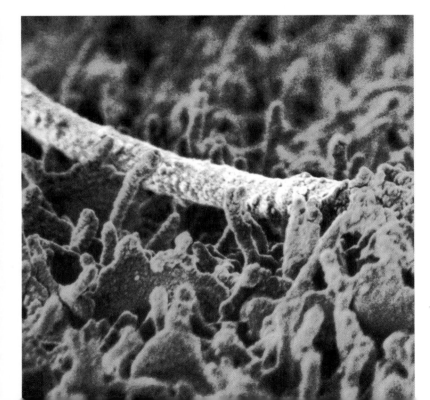

*First place, among millions of competitors. The victor is appointed!*

### The winner

The winning sperm has penetrated the innermost kernel of the egg, where the woman's genetic information is stored. The man's genes are packed tightly into the head of the sperm, behind which is the centriole that will play a key role when the fertilized egg is ready to begin dividing.

Soon the head will swell up, becoming a distinct nucleus. As a spaceship loses its rocket booster, the sperm loses its tail, as well as the expired "engines" in the midsection that have now served their purpose.

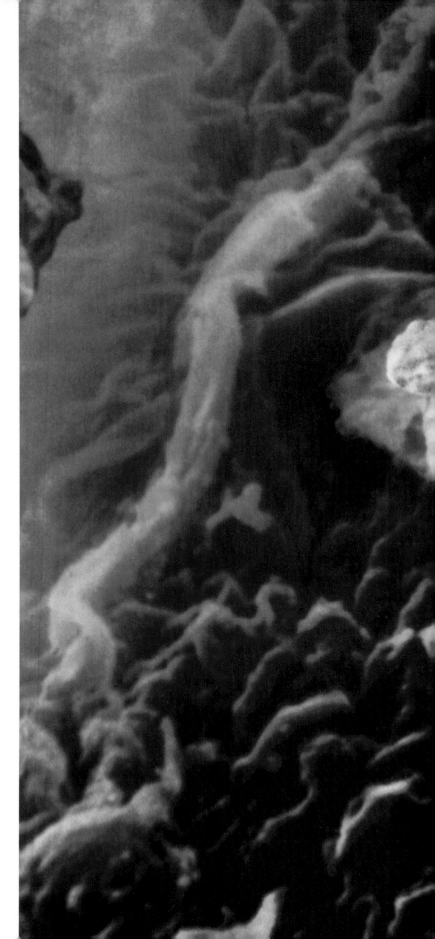

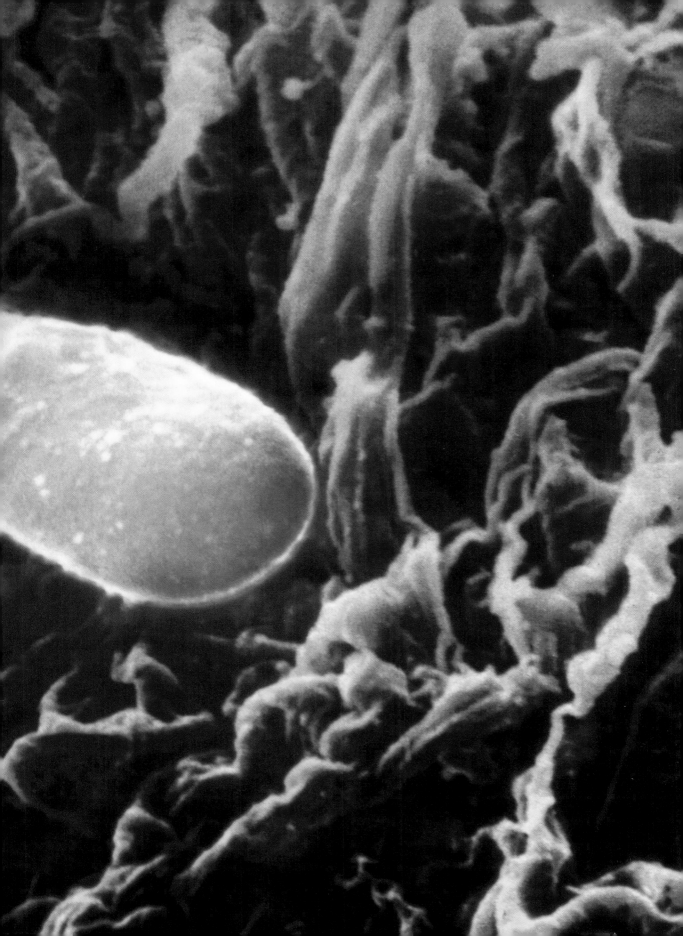

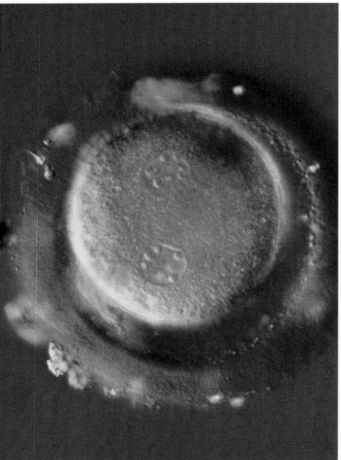

### Toward an enduring union

In the cytoplasm of the egg are now two cell nuclei of approximately the same size. One is the distended head of the sperm containing the man's genetic information; the other is the nucleus of the oocyte, containing the woman's genetic information (above left). They move slowly toward each other and toward the center of the egg (right).

## *Meeting deep inside the oocyte*

At about the same instant that the sperm-head swells up into a nucleus inside the cytoplasm, another nucleus is forming inside the egg, in the spot where all the genetic information from the woman has been stored. Now these two nuclei must meet. To help them, spiderweb-thin tubulin threads develop out of a part of the sperm-head called the centriole and link to the two nuclei. Today, using a microscope, we can watch the cytoplasm of the egg spin around, more or less kneading the nuclei and assisting their mutual approach. The threads serve as guide rails.

The nuclei are drawn inexorably toward each other, and when they meet, they fuse. At that moment a unique genetic code, a human embryo, is created.

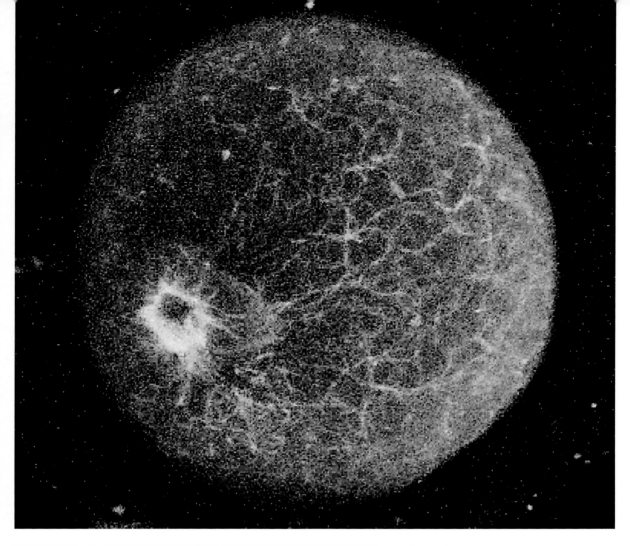

## Attraction

The power of attraction has its equivalent at the cellular level: the male nucleus attracts the female one with the aid of tubulin filaments, which crisscross the interior of the oocyte like a cobweb glimmering in the sun (above). The primary function of these filaments is to guide the material in the nucleus of the egg's cytoplasm toward fertilization, and to prevent fusion between the wrong pieces of genetic material.

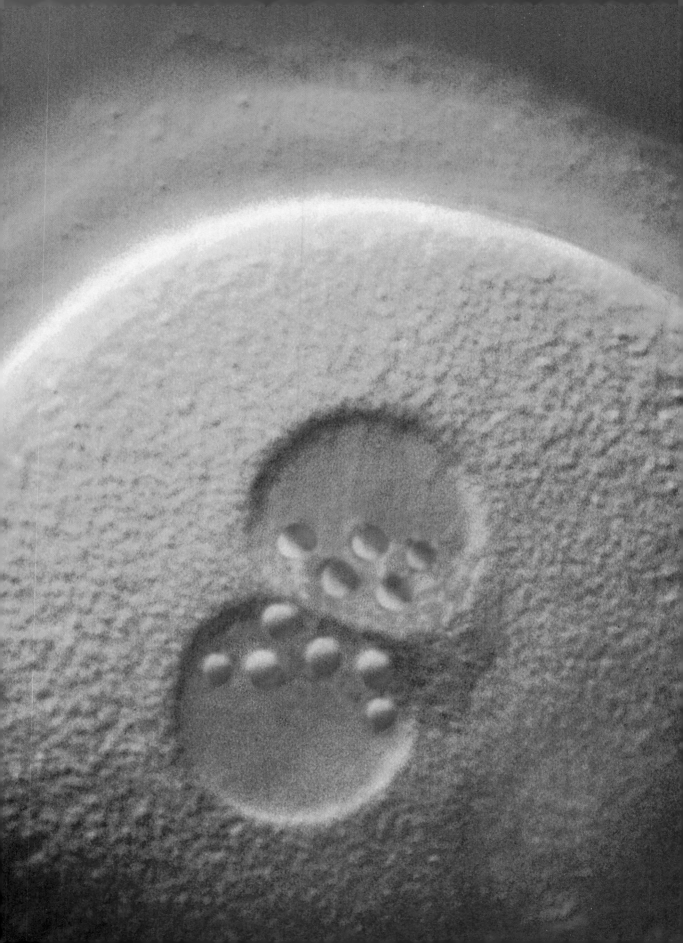

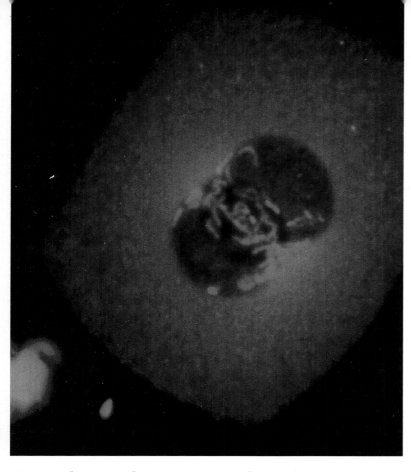

**Fusion**

Deep inside the egg, the miracle occurs: the nuclei of the male and female cells meet, and new chromosomes are created (opposite). At this instant a new, unique genetic code arises. The new individual is a product of its parents, with some genetic material from the mother and some from the father. The potential variety of combinations is virtually infinite. Many people define this as the moment life begins.

## A new human being is created

As soon as the sperm penetrates the egg's cell membrane, something amazing happens. The chemical composition of the egg undergoes a sudden change, as a rapid stream of ions modify the electric current across the its membrane. All other sperm are now excluded. This is essential, because if the egg were to be fertilized by more than one sperm, the genetic information would be disturbed, and development would stop.

After the nuclei fuse, their outer walls dissolve and they are incorporated into the cytoplasm of the egg. Fertilization is complete. Inside the eggshell there is now just a single cell, the original cell for all the billions of cells that will develop into the body of the future human being. Soon the cell will divide into two identical cells. The centriole, brought by the sperm, distributes the genetic material at this first cell division.

The uterine tube is perfectly adapted to the requirements of the newly fertilized egg (the zygote). Nutrients flow through the mucous membrane, and the tiny amounts of waste products that arise are diluted in the sea of fluids surrounding the egg.

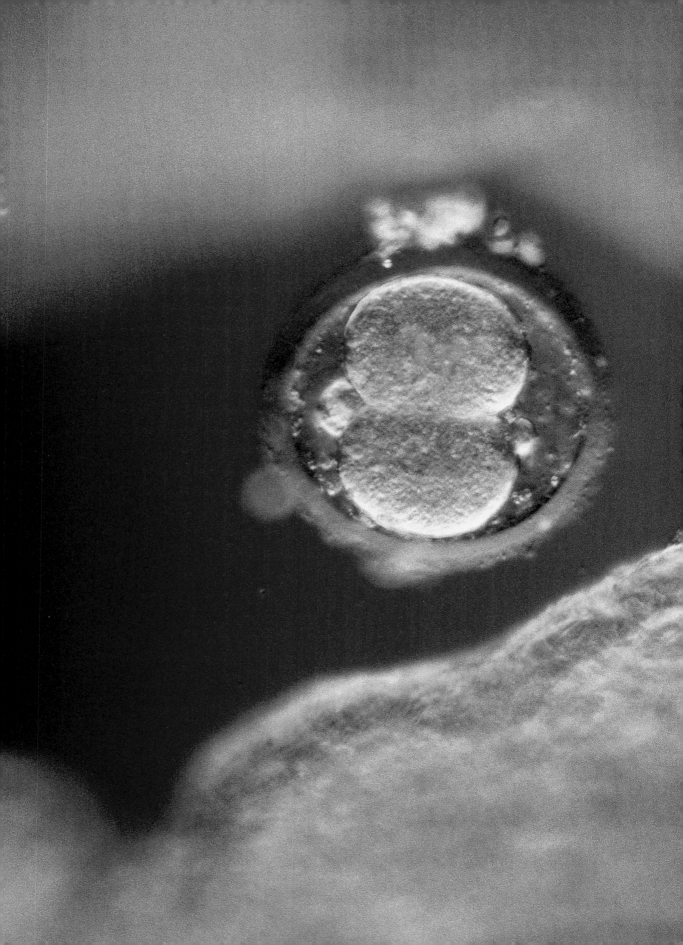

1 day

## The first cell division

While it is being fertilized, the egg rests in the folds of the mucous membrane of the Fallopian tube, rocking slowly back and forth, following the movements of the woman's body. A few days later it begins its journey down to the uterus; at the same time the single cell divides for the first time, creating a precise copy of itself. Such cell division is the cornerstone of all life in the universe, the key to the miracle of creation.

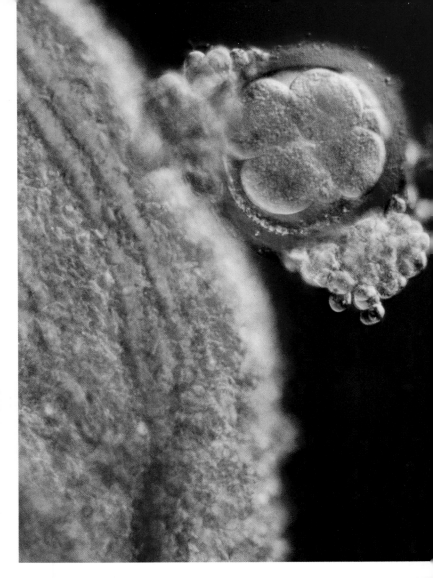

### Assisted transport

The fertilized egg is pushed gently forward by the tiny cilia covering the mucous membrane of the Fallopian tube, all swaying in the same direction (below right). This movement is reinforced by rhythmic contractions in the tube's muscles. Now the egg is composed of four to eight cells, and remains surrounded by some of the nutrient cells (above right).

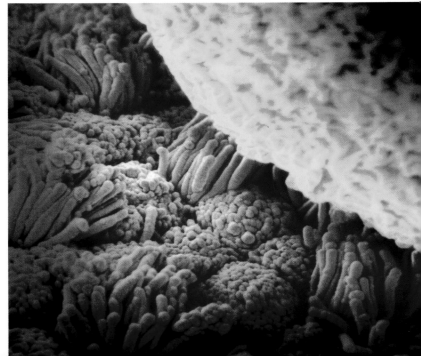

## *Journeying through the uterine tube*

The fertilized egg travels through the uterine tube for another two or three days, dividing several times, initially at twelve- to fifteen-hour intervals. Forty-eight hours after fertilization, there are four cells, and twenty-four hours later, eight. At this point it is very difficult to distinguish the individual cells, and in the fourth day the fertilized egg (the morula) resembles a mulberry. After another day and night, a hollow becomes visible, after which the morula is referred to as a blastocyst. At this point the first clear division of tasks among the cells can be detected. Some will develop into the embryo, while others will become the placenta.

On the mucous membrane of the uterine tube are millions of tiny hairlike cilia, all swaying in chorus in the direction of the uterus. The wall of the tube consists of muscles that can contract, conducting the flow of fluids in the same direction. In other words, the Fallopian tube is constructed to prevent an egg, be it unfertilized or fertilized, from accidentally slipping into the abdominal cavity.

The uterine tube narrows into a structure reminiscent of the lock system in a maritime canal, consisting of a muscle surrounding the tube. It has been tightly shut until now, but on the fourth or fifth day after ovulation the lock suddenly opens to allow the little blastocyst to pass through. The tension in this muscle and its consequent ability to open and close are regulated primarily by ovarian progesterone.

**The losers**

Outside the egg, sperm are still struggling valiantly, whipping their tails in vain, in hopes of fertilizing the egg. But now that there is a winner, the egg has effectively sealed itself off, preventing additional sperm from entering. These sperm may take up to several days to give up the struggle and die.

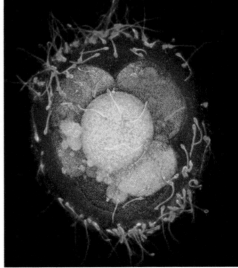

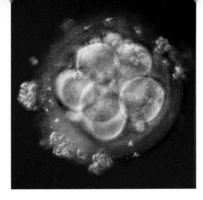

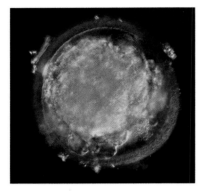

## Arrival at the uterus

This journey through the last narrow part of the tube before the uterine cavity takes only a few hours. But the blastocyst has to work its way through the folds of the mucous membrane, and it is a very tight fit indeed. Still, the job is a matter of life and death. Should the embryo get stuck and adhere to the lining of the tube instead of to that of the womb, the result will be an ectopic pregnancy, in which the blastocyst implants in this narrow part of the tube. After a few weeks the blood vessels that have established contact between the placenta and the tubal wall may burst, which can be terribly painful. The woman begins to lose blood into her abdominal cavity. In the worst case, her blood pressure falls and she faints, requiring emergency surgery.

An ectopic pregnancy hardly ever goes to term, though the little embryo tends to be perfectly well developed. Today most ectopic pregnancies are treated using endoscopic microsurgery. Not all ectopic pregnancies are dramatic. Some do not even require surgery but are treated medically or not at all, because the body released the blastocyst itself and heals normally.

**Cell multiplication**

Four days after fertilization the clump of cells has reached the morula stage and will soon contain twenty-five to thirty cells (top). Twenty-four hours later it has developed into a blastocyst with seventy to one hundred cells (above).

**Ready to exit the Fallopian tube**

The tiny blastocyst has made its way through the swaying landscape of the Fallopian tube and is now slowly approaching the narrow opening of the uterus (opposite). First, however, the cell divides into two groups inside its thin shell. One will become the new human being, while the other will be the placenta. In the photo to the right, the embryo can be seen at the upper left of the egg.

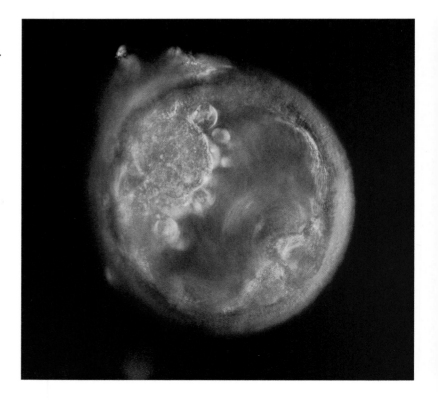

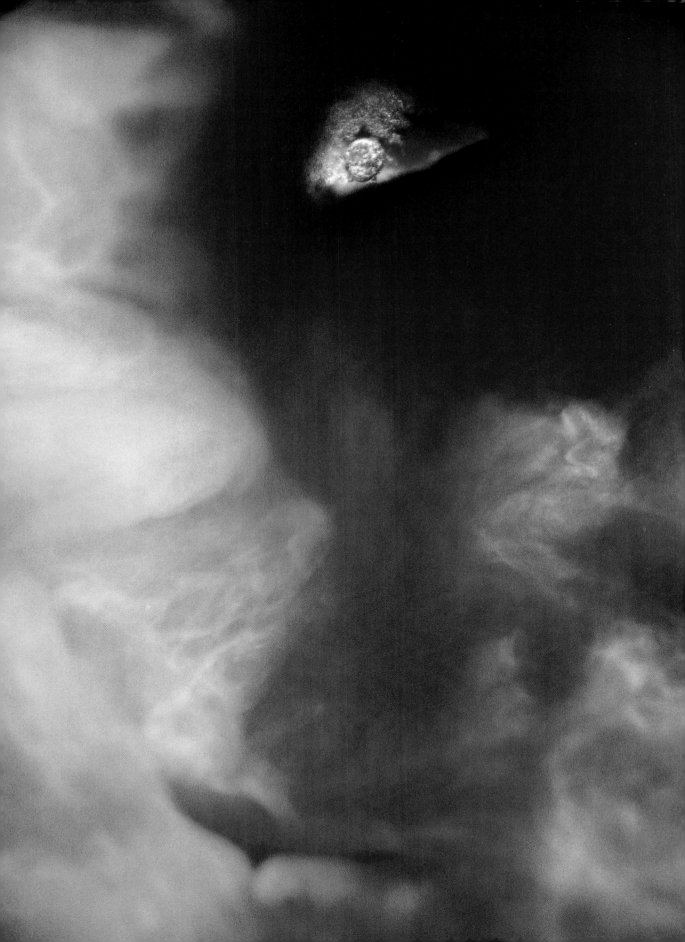

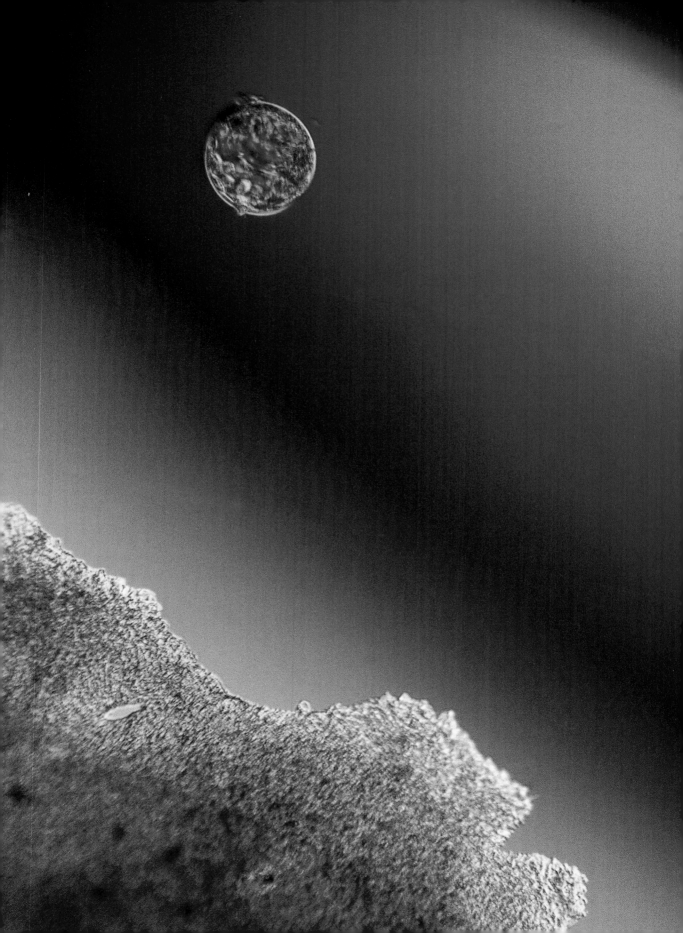

## In the uterus now

The trip through the uterine tube is over, and the blastocyst is now in the spacious womb. Here conditions are anything but crowded, and it can simply find a place to nest for nine months and grow. The uterus is well prepared for it. The mucous membranes have had nearly a week to develop since ovulation, while the egg was being released from the follicle, fertilized in the uterine tube, and transported to the womb.

Immunologically, the tiny embryo that will develop in the uterus is actually an invader, with a protein composition entirely alien to the mother and therefore subject to rejection. Normally, however, sophisticated systems in the woman's body ensure that the embryo remains and develops without being rejected. Very occasionally these systems malfunction, the immune system emits warning signals from the uterus, and the embryo is unable to grow. The woman may be perfectly capable of becoming pregnant, but she may experience repeated miscarriages, known as habitual spontaneous abortions. Today medical science can offer a woman with this problem special treatment to restore her body's ability to accept rather than reject an embryo.

**Close contact**

This woman and man are going to become a family, but they don't know it yet. Deep inside the woman's body, there is a tiny human embryo. Shortly, it will attempt to adhere to the wall of the uterus (opposite).

## Hatching

The embryo seeks a suitable nesting place in the uterus (right), but before it can nestle, it must shed its enveloping shell. It is the embryonic cells that are responsible for the adhesion to the mucous membrane.

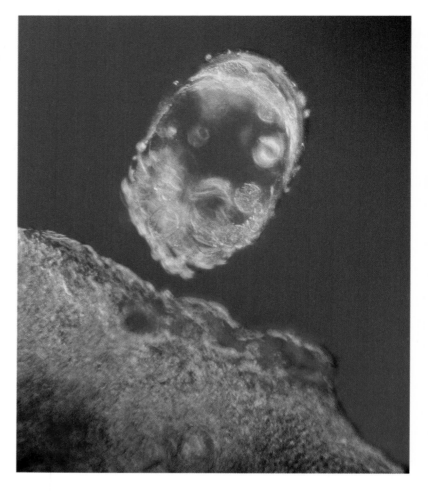

## *The eggshell breaks*

Just before it lands gently in the uterus, the embryo contracts and expands at least three or four times, then sheds its capsule, in a kind of hatching process. This "struggle for liberation" can be followed on film. If the little embryo is healthy and viable, it will hatch successfully, and the transparent, empty shell can be seen sailing off and dissolving.

While the outside of the capsule was relatively smooth and hard, the new surface of the egg is more rippled, and stickier. It is as if the entire embryo has been dipped in sugar solution. Shoots of sugarlike molecules reach out for the surface of the uterine mucous membrane, which, in its turn, has similar sugarlike molecules into which the little shoots fit.

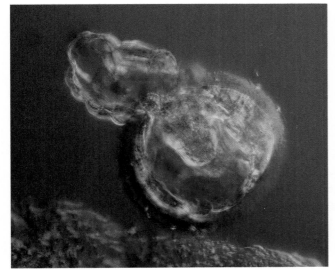

## Liberation

During the hatching process, the blastocyst tumbles around inside the uterus. Now and then it bangs into the soft mucous membrane lining.

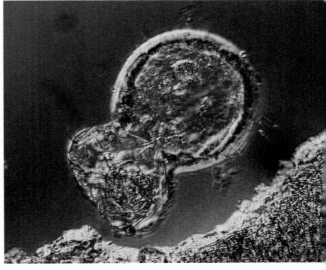

## Contact

The mucous membrane embraces the fragile, newly hatched cell mass (below). Usually the embryo implants in the upper half of the womb.

**7 days**

## The seventh day

Once the shell has broken, the embryo expands
extremely rapidly. This friendly intruder gradually
settles into the mucous membrane of the uterus.

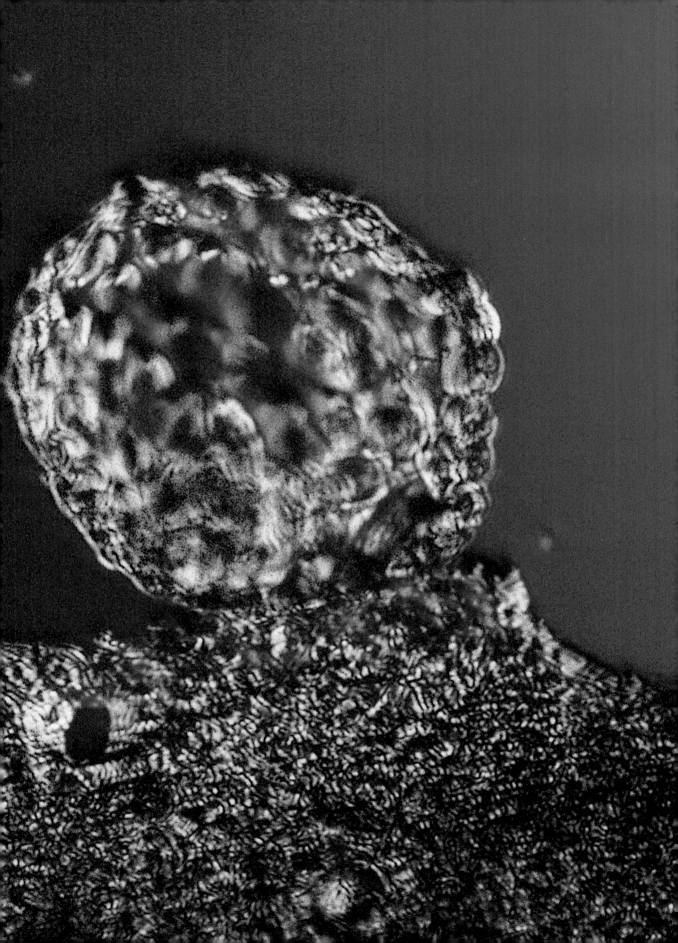

## The embryo has landed

The little embryo seems to select the spot in the uterus into which it will settle. It appears to emit chemical signals to its surroundings, which respond by offering a good environment for development and growth. Today a great deal of research is focused on determining what makes a particular spot a good environment, and what factors are most important during this implantation phase. This research should make it possible to improve the techniques used for artificial fertilization, thus increasing the chances of pregnancy for many women who are otherwise infertile.

The moment of first contact between the embryo and the uterine lining is critical. At this one instant, for implantation to succeed, many factors must come together precisely. Shortly after the landing, contact intensifies. The cells that will form the placenta penetrate the mucous membrane, starting a dynamic exchange of chemical signals, nutrients, and oxygen. Tiny blood vessels that grow in through the mucous membrane receive hormonal signals from the developing placenta. These signals inform all the systems in the woman's body that a new being has begun to grow in her womb.

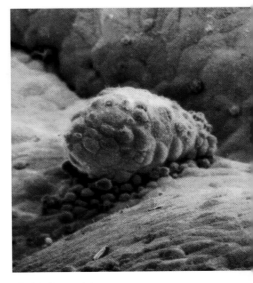

**Eight days old**

The tiny embryo has now adhered to the moonscapelike lining of the uterus (opposite). The choice of implantation site was not a random one. Here and there in the mucous membrane little hillocks of rounded structures took shape; they emitted chemical signals that apparently functioned as inviting landing beacons, without which implantation—the first contact between the embryo and the mother-to-be—cannot take place (above).

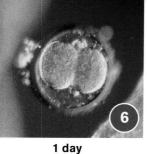

### Cell division

Deep in the Fallopian tube the original cell divides, becoming two precisely identical cells.

**1 day**

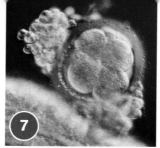

### Four cells

The fertilized egg, still surrounded by nutrient cells, on its way through the Fallopian tube.

**2 days**

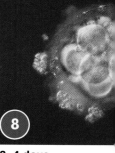

**3–4 days**

**18 hours**

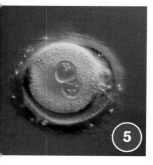

### Fusion

The nuclei of the sperm and the egg meet. Their genetic material combines, giving rise to a new human being.

**10 hours**

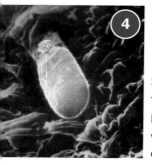

### Success

The winning sperm has now made its way to the interior of the oocyte.

### Ovulation

A large ovarian follicle on the surface of the ovary releases a mature egg, carefully prepared for fertilization.

**6 hours**

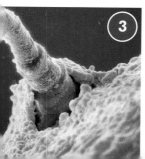

### Last lap

The sperm have reached the egg and are trying to penetrate the shell (right)—the winner may be visible on the left.

**5 hours**

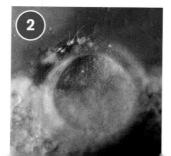

**1 hour**

### Waiting

The egg, surrounded by nutrient cells, in the outer part of the Fallopian tube.

### Expanding berry

The fertilized egg (morula) looks like a mulberry. All the cells are still the same size.

**5 days**

### Tight squeeze

The fertilized egg navigates a narrow part of the Fallopian tube just before reaching the uterus.

### In the uterus

The embryo leaves its shell and tries to adhere to the wall of the uterus.

**6–7 days**

**9**

**10**

**8 days**
### Implantation

The landing at the chosen site has been successful. The surface of the uterus is here and there covered with pinopodes, which play an important role in making contact between the cells of the embryo and those of the mother.

## The first few days

From the moment of ovulation until the moment when the fertilized egg adheres to the lining of the uterus, about eight days have passed. The entire process is completed in the wide outer reaches of the uterine tube, known as the ampulla, where the egg remains for about three days before migrating toward the uterus. Throughout the journey the fertilized egg retains its protective shell, but upon its arrival the shell must break to enable the sticky surface of the embryo to attach to the mucous membrane of the uterus.

# Pregnancy

In the darkness of the uterus, over a period of nine months, a cluster of cells is gradually transformed into a new little person, ready to encounter the world. For expectant parents, this is a period of waiting and anticipation, of concern and joy. Unbelievable–we're going to have a baby! Will everything go well? Will it be a boy or a girl?

# The first few weeks

## The pregnancy timeline

A woman with a twenty-eight-day interval between her periods will ovulate on the thirteenth or fourteenth day. If she has intercourse while she is ovulating, the egg will probably be fertilized by a sperm within twenty-four hours of ovulation, after which the fertilized egg will take just over a week to make its way through the Fallopian tube and implant itself into the uterine lining. Once this has happened, the woman is pregnant. The pregnancy timeline is usually measured not from the date of ovulation or from the date of intercourse, but from the first day of the woman's last period. On this timeline a pregnancy lasts forty weeks, in spite of the fact that in a full-term pregnancy the embryo will spend only thirty-eight weeks in the mother's Fallopian tube and uterus.

In this chapter, unless otherwise noted, the length of pregnancy is counted from the date of the woman's last period. In this section, "The first few weeks," the actual age of the embryo is also sometimes used.

## *Early signs*

When a woman becomes pregnant, her usual hormonal cycle begins to change. Just a week or so after conception, she may become aware that something is different. Her breasts will become tender, and she will be more sensitive than usual to smells. When her period fails to begin, this will help confirm something she already suspects. These early signs are brought about by changes in her hormone balance, not by the tiny new coagulation of cells in her uterus. She will probably feel very tired early in pregnancy, exhaustion that can hardly be caused by that little clump in itself.

Even before it implants, the clump will have divided into two distinct parts; one known as the inner cell mass, will become the embryo, and the other the placenta. The latter cells attach themselves firmly to the uterine lining, and some turn into blood vessels that begin to communicate with blood vessels in the wall of the womb. In this way, the mother's blood will provide the embryo with all the nutrition and oxygen it needs and allow the embryo to rid itself of waste products from metabolism.

Some of the cells in the developing placenta give rise to an important hormone, human chorionic gonadotrophin (hCG), that signals to the ovaries and the pituitary gland that the woman is pregnant. These signals make it clear that there will be no need for ovulation for quite a while and that the uterine lining should not be expelled—that is: there should be no menstruation. The corpus luteum in the ovary responds to these instructions by creating more progesterone, which reaches the uterus via the circulatory system. Progesterone is the hormone needed by the mucous membrane of the uterus in order to grow and provide the embryo with the environment it needs.

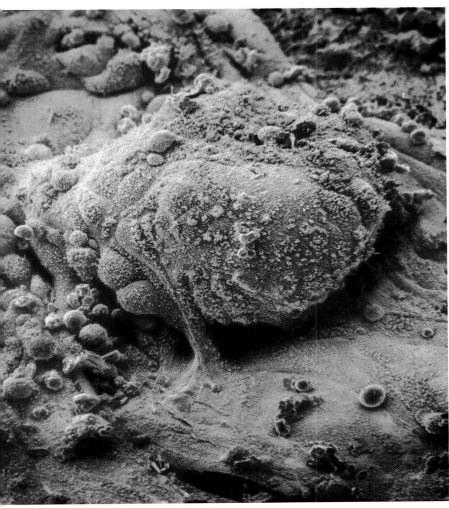

## Premonitions

As early as about ten days after fertilization, the level of progesterone (above) in the blood rises dramatically. Many women feel this change distinctly. A woman's breasts become tender, even more so than prior to menstruation, and some morning nausea is not unusual. Inside the womb the embryo has just been implanted in the uterine lining, entering into an intimate alliance with the mother-to-be.

The embryo in this photo (left) is ten days old.

**A fraction of an inch long**

At the beginning of week 5 of pregnancy, the embryo is just a few millimeters (a tenth of an inch) long, and its curved body is soft and transparent. Along the length of its body runs a nerve tube, and the brain has just begun to form at the head end (right).

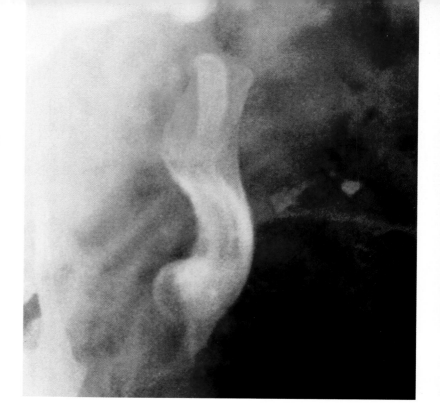

*The 22-day-old embryo does not yet have a face, and the brain is unprotected and open.*

## A human tadpole

Just a week or so after implantation, the formless clump of cells becomes oval and then oblong and begins to look vaguely like a worm. For this to happen, the embryo must have established firm contact with the mother's bloodstream via the placenta. Many embryos fail to do so, and a miscarriage ensues.

Even this early a superficial layer of cells outlining the embryo is discernible. This layer is called the outermost germ layer, and over the course of a few days it thickens around the midline of the body, forming two lengthwise folds. Between these folds an indentation deepens, then quickly closes, making a tube. At one end the primitive brain begins to form like a little bubble. Nerve fibers begin to protrude from the brain stem, and the spinal cord starts to form. On approximately the fifteenth day in the life of the embryo, the first primitive nerve cells, which in time will govern bodily functions and provide the spark of consciousness, are formed. Some consider this to be the point when life begins, since from these nerve cells the brain and consciousness will eventually be built. Without nerve cells there would be neither expression nor impression, and thus the emergence and death of the brain can be seen as the beginning and the end of life.

week
5

## Rapid, daily changes

The formless little clump of cells has been attached to the lining of the uterus since some eight days after fertilization. At this point the inner cell mass, the child-to-be, consists of several hundred cells, all of which contain the same genetic code. But no cell can express the entire code at once, and only part of it is expressed in each cell.

As early as the division that distinguishes placenta cells from embryonic cells, the potential of each individual cell is radically restricted. Those in the outermost germ layer will go on to become the brain, spinal cord, and nerves, as well as skin, hair, sebaceous glands, and sweat glands. The cells in the middle germ layer will develop into the skeleton, the heart muscle wall, and the other muscles. They will also become blood vessels, lymphatic vessels, and blood corpuscles that, together with the heart, will form the circulatory system. The ovaries and testicles, as well as the kidneys, will also come from this layer. The inner germ layer will give rise to the intestinal system—the stomach and the small and large intestines—and the urinary tract. This layer will also provide the mucous membranes of the entire body as well as the lungs, an organ that will not begin to play a major role until birth.

This early stage is an extremely sensitive one, as even the slightest deviation from the program may result in damage and deformities that the child-to-be will carry throughout life. During this time both the external and the internal environment are incredibly important. Smoking and drinking are only two of the factors that pose a risk to the normal development of the embryo (see page 119).

Not only the individual organs but also the communication between them has to develop perfectly, down to the minutest detail. If a serious defect arises, natural protective mechanisms go to work, and the woman has a miscarriage. Although miscarriages may have other causes, genetic abnormalities are the most common. Nearly one pregnancy in five comes to an end at this stage, because either the egg or the sperm lacked the optimum genetic prerequisites. If a woman develops bleeding and/or abdominal pain at this stage, she should consult her physician, the hospital, or the prenatal clinic so that the reasons can be investigated. The beginning of a miscarriage is one possible explanation, but there are other kinds of bleeding that do not threaten the pregnancy.

**17 days approx.**       **20 days approx.**

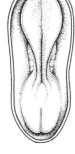

## From a cluster of cells to an embryo

When week 5 begins, the embryo changes rapidly. Within just a few days it is transformed from a clump of cells into an oblong body, with a head and a tail beginning to take shape around the neural tube.

head with the beginnings of the forebrain

the nerve tube has closed at the torso end

the primitive heart

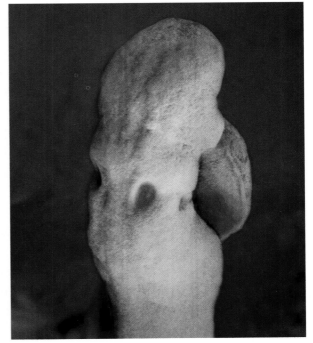

**A human being– age 22 days**

The backbone of the embryo is curved, with the neural tube open at both ends of the body (left). The outer germ layer, the skin of the embryo, now encloses the torso.

**23 days**

The embryo has begun to straighten, though now it seems bent over the immense heart (above). The head is still wide open, and we can just discern the brain cells.

**25 days**

The skull bones have begun to fuse, and primitive gills are bulging slightly on the otherwise smooth throat (above). The hollows toward the back of the head will become ears.

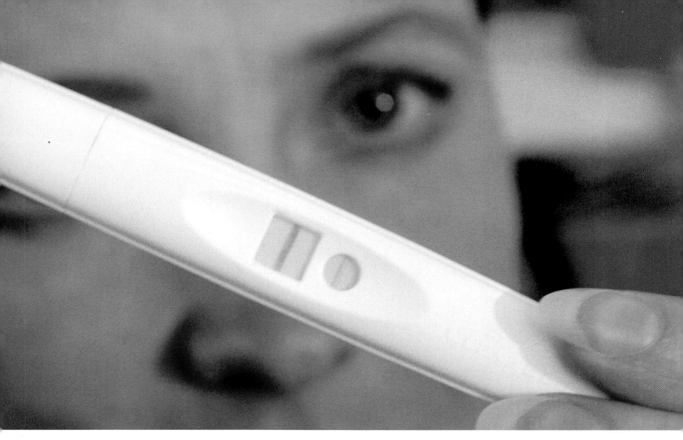

**Blue line = pregnant**

Today it is easy to determine whether a pregnancy has begun. All it takes is a plastic stick with a color indicator. The stick is dipped in a urine sample; if the color changes, the woman is very probably pregnant (above). This urine test can be used just a couple of days after a missed period, and it works thanks to a hormone that develops very quickly after fertilization. This hormone, human chorionic gonadotrophin (hCG), enters the woman's circulatory system and can therefore be traced in her urine after a few days. A pregnancy test, in the form of a urine or blood test, can also be performed at a prenatal clinic.

## Confirmed!

Today there are simple, sensitive tests a woman can use to verify whether she is pregnant as early as her first missed period and sometimes even earlier. The interplay between the embryo's hCG and her own progesterone may make her feel that something has happened. Human CG is a hormone unique to pregnancy, as it can be produced only in placental cells. When this hormone is filtered out of the bloodstream via the kidneys, a urine sample may reveal the pregnancy.

When will the baby be born? Although there is no precise length of pregnancy, a woman will usually be given an estimated date, for both medical and psychological reasons. This date is based on an "ideal" pregnancy, lasting forty weeks from the last menstrual period. One traditional method of estimating the day the baby will be born is to count from the first day of the last period, subtract three months, and add seven days. This will work only if the woman has had regular periods at twenty-eight-day intervals. If the intervals are longer or are irregular, calculating the due date will be far more difficult. Other methods long used by doctors and midwives include measuring the size of the uterus

**We're going to have a baby!**

The little line on the pregnancy test stick looks quite unassuming, but it is likely to trigger strong feelings. It means that inside the woman an embryo is growing, and dramatic changes are under way.

when the woman comes to the clinic and counting from the date when the woman notices the first fetal movements. In recent years ultrasound has come to be considered the most reliable way of dating the age of the fetus and the delivery date.

Most women who become pregnant are thrilled, and the idea of a baby is a very welcome one, but other women must, for one reason or another, make the difficult choice of terminating the pregnancy. Statistics indicate clearly that very young women under the age of twenty and women over the age of forty are those who most frequently decide to terminate. The number of abortions varies from country to country and depends on the legality of abortion and the availability of contraceptives. In the United States and the Nordic countries a woman has the right to terminate her pregnancy in its early stages. In other countries, such as Ireland, the laws prohibit abortions, and a woman must undergo the procedure illegally, which has serious risks. A poorly performed abortion can result in future sterility because of infection. Globally the most common cause of death in women between the ages of fifteen and forty is complications after an illegal abortion.

## The first heartbeats

The heart begins to develop when the embryo is still but a cluster of cells, and as early as its twenty-second day the newly formed heart muscle cells contract, and the heart beats for the very first time. The mission of the heart, central to the development of the embryo, is the circulation of blood, which distributes nutrition and oxygen to all the tiny developing organs.

At this stage the heart already has two chambers (ventricles) and is so large that it almost seems to be outside the rest of the body. The right-hand chamber takes up the blood from the other body organs, and the left-hand one releases freshly oxygenated blood to the body. After birth, blood from the body will return to the right-hand chamber, which will pump it into the lungs to pick up oxygen. Then blood will flow into the left-hand chamber, where it will be pumped into the aorta, the large blood vessel that distributes blood throughout the body. In the fetus, however, the lungs are collapsed. The placenta oxygenates the blood, which returns to the heart to be directed to the rest of the body. At birth the lungs will expand and take over from the placenta. The heartbeat of the embryo is very rapid, nearly twice that of the mother, and can easily be heard even with very simple listening devices. Heart rate is one of the most reliable ways of knowing how the fetus is faring.

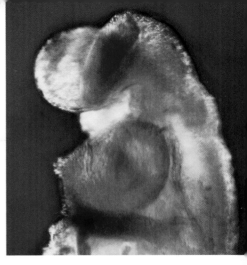

**Heart in mouth**

The primitive heart, which accounts for much of the body of the embryo, has been still until now, but on the twenty-second day after fertilization, it suddenly contracts and begins to beat (above). Two days later the heart is ticking fast and rhythmically (opposite).

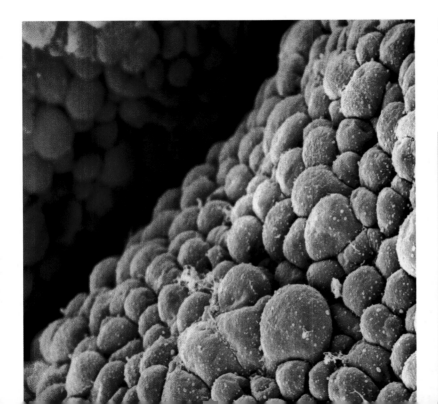

**Heart muscle at work**

When one muscle cell contracts, there is a domino effect on the surrounding ones. No separate heart nerves yet govern the heartbeat.

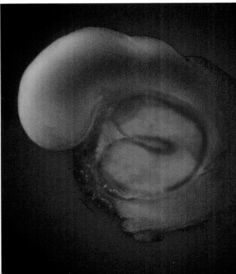

**week 5**

## The world of the embryo

The tiny embryo is securely implanted in the uterine wall. The body has the shape of a seahorse, and the little heart extends almost all the way up to the head. The placenta, with its branching vascular network, is the link between embryo and uterus.

**24 days, 4 mm./0.15 in.**

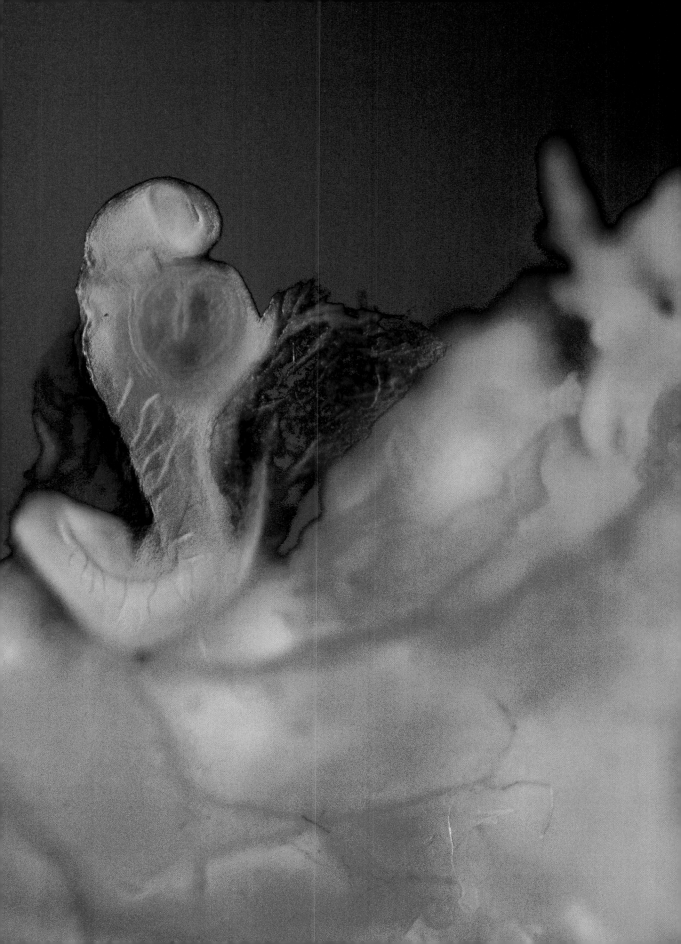

## Early back view

The embryo's back is turned toward us (above), and we see the whole vertebral column, running down from the neck to where the legs will be. The arm buds extend like little wings. The placenta is much larger than the embryo at this point. Suspended on the left, hanging like a balloon, is the yolk sac, which among other things may influence the development of the genitals.

**28 days, 6 mm./0.23 in.**

## *A vertebrate—but what kind?*

Even in the embryonic stage, human beings are clearly vertebrates. On both sides of the trench of nerves along the backbone, skeletal building blocks form from the middle germ layer. These develop into thirty-three or thirty-four vertebrae, although the four caudal vertebrae, left from a past stage of evolution, fuse into a single bone, the coccyx or tailbone. Those closest to the head are called cervical vertebrae.

Between the primitive opening that will become the baby's mouth and the bulging heart sac, seven shoots develop from the vertebral column. At first these shoots look like fish gills. Soon, however, one of them grows into the baby's jaw, while the others become the face and neck. Below these the twelve thoracic vertebrae are formed. The ribs grow out of these vertebrae, arching to shape the baby's chest. Inside this cavity there are already primitive lungs. The vertebrae must not fuse—if they did, the backbone could not bend. Elastic tissue and muscles will hold the vertebrae together and gradually steady the backbone.

From little hollows between the vertebrae, tiny bundles of nerves reach out and spread a delicate web through the body. This network will play two very different and important functions: the brain and the spinal cord will emit signals to all the muscles in the body, instructing them to contract and perform various motions; and information will be returned to the brain via the spinal cord, informing the brain of what the rest of the body is doing. This signaling system begins to operate fully when the embryo is six or seven weeks old. Some nerves register touch, while others register pressure, temperature, and so on. Other special nerve impulses come to the brain from the eyes and nose and from the mouth and tongue. Thus an entire nerve structure serving our senses is constructed very early in life indeed.

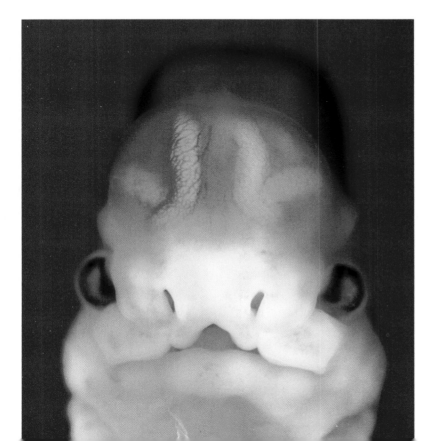

### Face to face

Now we can see the eyes, nose and mouth—a developing face (left). The nostrils have formed above the opening that will be the mouth, but the groove between the nostrils and the edge of the top lip has not yet been effaced. The little hollows that will be the eyes are almost all the way out by the temples.

At this early stage fetuses developing in the uteri of a pregnant ape, pig, and human being all look very much the same, but during the coming weeks human features become more and more prominent.

**30 days, 7 mm/0.26 in.**

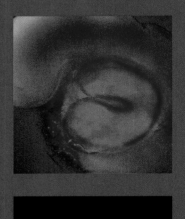

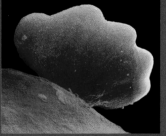

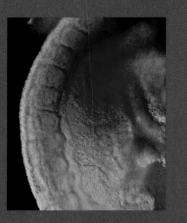

# The embryo takes shape

The embryo is now five weeks old and well past the stage when it looks like a formless clump of cells. The skin layers are still barely developed, and the tiny body is quite transparent. The head and tail can be distinguished, as well as the heart, the vertebrae of the spinal column, and the beginnings of a tiny hand.

### ◀ The heart

The heart has been beating rapidly for a couple of weeks now, pumping blood to tissues throughout the body.

### ◀ The hands

For the time being, the hands resemble paddles. The individual fingers are just barely discernible (see page 142).

### ◀ The vertebral column

The vertebral column or backbone gives the body stability and houses the spinal cord from which nerve fibers extend into the rest of the body. This detail shows the vertebrae.

### The eyes

Although the eyes have begun to take shape, there are as yet no eyelids (see page 146).

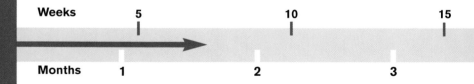

| Weeks | 5 | 10 | 15 |
|---|---|---|---|

| Months | 1 | 2 | 3 |
|---|---|---|---|

## Approximate size

Length 5–10 mm./0.19–0.39 in. (The measurement is given as crown rump length, or CRL: the length of the embryo from head to bottom, in a sitting position.)

## Tiny and tenuous

So far so good, but there is still a substantial risk of miscarriage.

## The brain

The top section of the spinal cord has swollen to a little rounded bump, the primitive brain. Even now myriads of couplings between nerve cells and different parts of the brain are developing, to take on their separate functions (see page 106).

## The mucus plug

As soon as fertilization has taken place, a plug of mucus closes off the cervix, keeping both bacteria and other sperm from entering the uterus.

## Doubling in size in a week

The embryo is growing fast. With every day that passes, it grows about 1 millimeter (0.039 inch). During the seventh week the embryo more than doubles, from about 5 millimeters (0.19 inch) to about 10 millimeters. (Length can be measured with ultrasound, which can also register the heartbeat at this early stage.)

An ultrasound image shows the embryo at rest toward the lower edge of the uterine cavity (right).

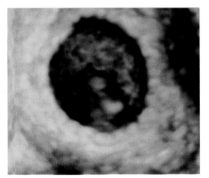

| 20 | | 25 | | 30 | | 35 | | 40 |

| 5 | | 6 | | 7 | | 8 | | 9 |

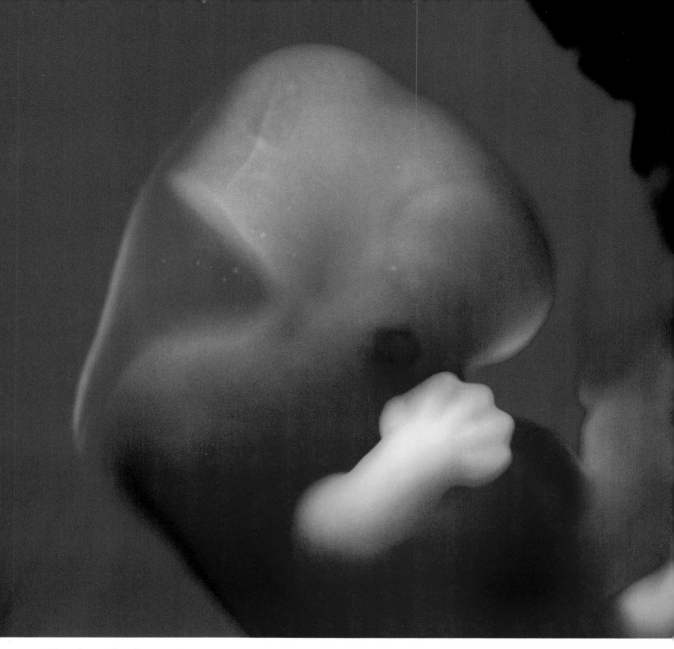

## Top-down development

In week 8 the head still dominates, and the upper body is much larger than the lower half. The arms and hands have progressed more than the legs and feet, but there is still only a suggestion of fingers.

**39 days, 12 mm./0.4 in.**

## *A little human begins to take shape*

When the embryo is about six weeks old, its appearance begins to change. No longer resembling the embryo of any primitive mammal, it now begins to look like a miniature human being. The head, hitherto tipped far forward, straightens up. The skull bones are not yet very ossified, so the brain shows right through the thin cartilage. In the forehead area big bubblelike structures will become the cerebrum, and behind it three smaller ones will eventually build up other important parts of the brain.

The head is quite huge in relation to the rest of the body, because the growth of the embryo takes place from the head downward. Not until adolescence will the body completely catch up, and in a newborn the head still accounts for about one quarter of the body's total length.

Many of the organs have already begun to function: the kidneys produce urine, and the stomach produces gastric juice. The embryo has also begun to move: the first "visible" motion is the rapid, steady beating of the heart, but soon small bodily movements show that nerve impulses coming from the brain are instructing muscles to contract. These begin as global motions, affecting the whole body, but gradually specific little movements take place, such as one hand moving while the rest of the body is still. This constant motion is important, stimulating normal growth and development of the muscles and joints. Should an arm or a leg be locked in position—by the umbilical cord, for example, or by a rupture in the amniotic sac—development of that body part may be disrupted, and it can atrophy. (This happens very seldom.)

### Jumping for joy

Even this early in pregnancy, the embryo is extremely lively, in constant motion, sleeping for only brief periods. The nerve fibers that govern motion and receive all the sensory impulses extend into the arms and legs.

The photographs below were taken using three-dimensional ultrasound technology and show even this early how the little embryo looks and its location and position in the uterus.

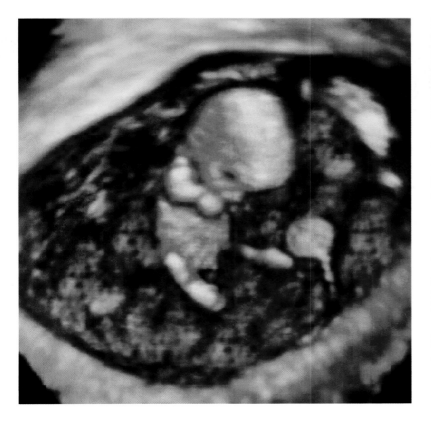

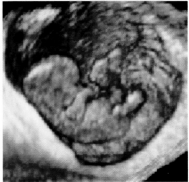

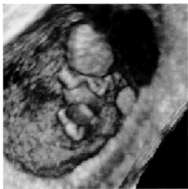

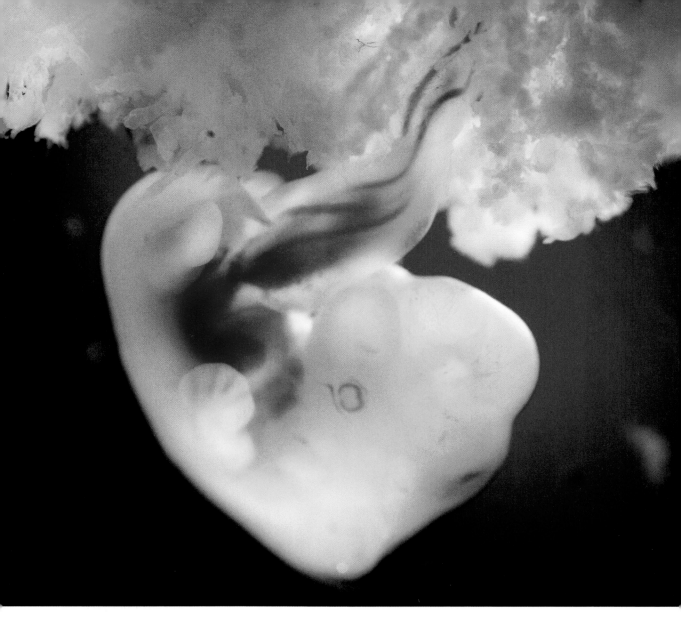

## The tree of life

The placenta is a sort of supply center from which the embryo gathers oxygen and nutrients and into which it discards waste products, throughout pregnancy, via the umbilical cord.

**40 days, 15 mm./0.59 in.**

## *The placenta provides oxygen and nutrition*

After implantation the placental cells continue to spread across the surface of the uterine lining, and grow down into its depths, coming into contact with the mother's circulatory system to assure the rapidly growing embryo a supply of nutrition and oxygen. The growth of the placenta keeps pace with that of the embryo and provides these vital supplies throughout pregnancy. The placental cells develop entirely out of the embryo and share its unique genetic code.

Early in pregnancy the placenta begins to produce progesterone, which remains a vital hormone throughout pregnancy. As

**Sophisticated exchange system**

Whatever food the mother consumes she shares with the embryo through her circulatory system. The placenta has a vast, branching vascular system (top). It also supplies the embryo with oxygen. Blue corpuscles, indicating oxygen-depleted blood, pass from the embryo through the placenta, to be reoxygenated via the mother's bloodstream. The returning corpuscles are red (above).

early as seven to eight weeks after the woman's last menstrual period, the placenta produces all the necessary hormones; the ovaries are no longer needed for that purpose. Hormone production by the placenta is essential for the normal continuation of pregnancy and for the fetus to develop "according to plan." Placental hormones will also be important in determining when the contractions of labor begin. As long as the progesterone level in the blood remains high or even rises further, the muscles of the uterine walls will remain relaxed. When these levels sink, the muscles will contract, and labor pains will begin.

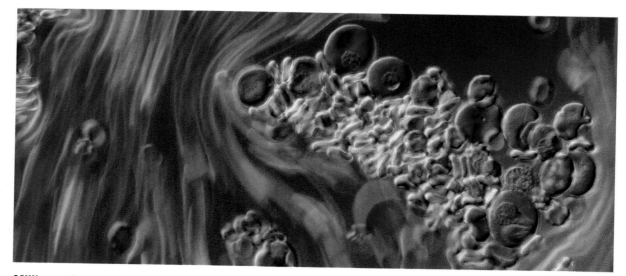

**Millions of corpuscles in motion**

In the tiny capillaries of the placenta where the exchange of nutrients from mother to embryo takes place, the blood corpuscles crowd tightly, waiting to be charged with oxygen and calorie-rich substances from the mother's blood. Once the corpuscles are recharged, they transport their load out to the organs of the embryo. During the fetal stage the corpuscles have a core, which vanishes after birth.

**The lifeline for the embryo**

Blood pulsates plentifully through the umbilical cord, transporting everything the embryo needs. The umbilical cord is short and thick at first, but it becomes longer, enabling the growing embryo to move more freely (opposite).

**46 days, 2 cm./0.8 in.**

## Circulatory systems meet

Early on the embryo develops a primitive circulatory system, which gradually enlarges to allow blood to circulate out through the aorta to the thin little capillaries in the organs and then back through the liver to the heart. Between the placenta and the embryo runs the umbilical cord, which contains three vessels: one big thick one that carries the oxygenated, nutrient-rich blood to the embryo's heart, and two smaller ones that carry the oxygen-poor waste-product-containing blood to the placenta. The two circulatory systems remain separate, although a small number of fetal cells do cross into the mother's circulation. Doctors use these cells and their DNA to gain genetic information about the fetus. Some substances that the mother consumes, including some medications, do not pass through the filter of the placenta and affect the mother only. Other substances pass readily through the placental filter and can hurt the embryo or fetus. Alcohol is, of course, the classic example.

Normally the fetus is not exposed to infection, since neither bacteria nor viruses pass through the placental filter. This protection is particularly important during the first half of pregnancy, when the fetus does not yet have a well-developed immune system and when its developing organs are more easily damaged. Although the fetus's immune system has begun to work by the fifth month of pregnancy, it will take a long time to be fully mature.

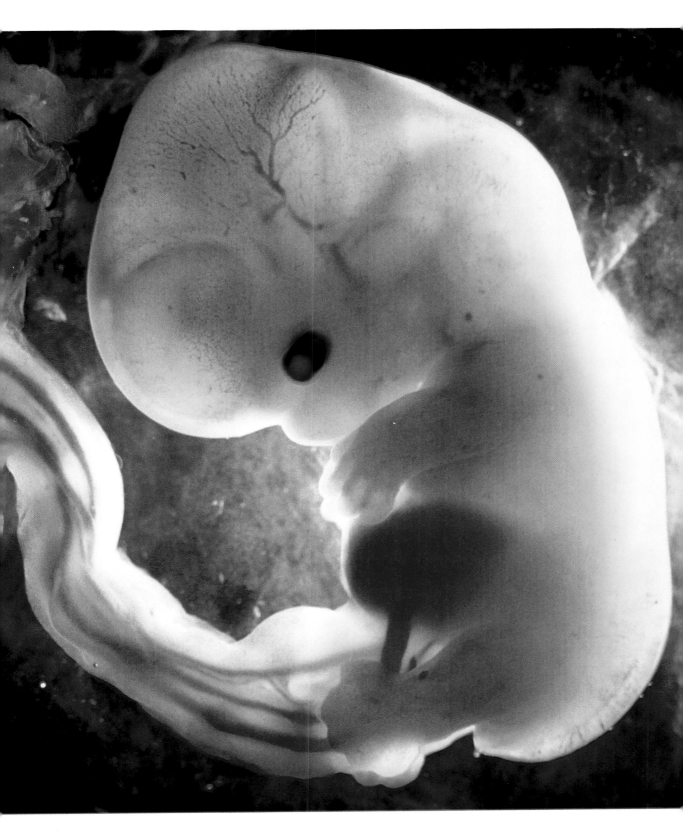

## A helmet of bone

At first the brain is wide open and unprotected, but soon this most important, most sensitive bodily organ is covered by the skull bones; these are only loosely joined, to leave space for the brain to grow. The head of an embryo in week 5 (above) and in week 10 (below).

## The early brain

Just a few weeks after fertilization, primitive nerve cells, initially almost round, are visible in a human embryo. Gradually, long runners known as axons develop from the bodies of these nerve cells. They establish contact either with other nerve cells or with an end station, such as a muscle, to which they will report information. Each nerve cell transports information mainly via electrical impulses, with a chemical transfer at the points of contact.

One of the substances responsible for this chemical transfer is noradrenaline, which rapidly relays information to the muscles and elsewhere. Dopamine and adrenaline are similar chemical transmitters. Still another chemical, acetylcholine, transfers information to the stomach and intestines but starts to work later in fetal life. Although the acetylcholine transport system is far less rapid, it is no less effective.

The nerves of the embryo, particularly while they are being established and developed, are extremely sensitive to toxins. Exposure to large amounts of alcohol, for example, at this early stage of development may cause permanent damage to the embryo.

The dynamic and stringently programmed development of the brain is essential for the growth of the body, as well as for the ability of the arms and legs gradually to begin to move in accordance with the normal pattern. The nerve stems, parallel bundles of axons, develop early and may be compared with broadband telecommunication, transporting complex information throughout the body at great speed. After some time the different parts of the brain become more clearly differentiated and take on their separate tasks. Some parts of the brain receive only sensory impressions from the body, such as sensation and pain, while others are responsible for vision and hearing, and still others govern movements, which begin awkwardly and gradually become better coordinated and more purpose-oriented.

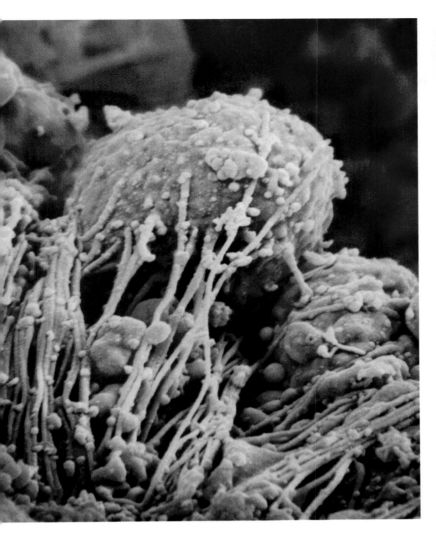

## Primitive nerve cells

Brain cells from a twenty-three-day-old embryo. Little outgrowths (axons and dendrites) have begun to protrude and are striving to make contact with cells in their proximity. A nerve cell receives a bunch of dendrites, delivering information (left). These nerve cells (above) are greatly magnified.

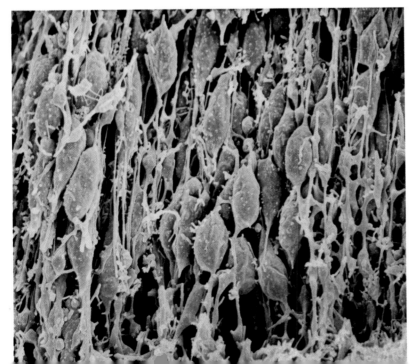

## Increased exchange of information

The nerve cells in the brain of an eight-week-old embryo are similar in construction to an old-fashioned telephone switchboard, but already they function far better, and their capacity is incomparably greater (left). More and more systems are becoming interlinked and are beginning to communicate. Some issue commands, while others receive signals and do the work.

# The embryo becomes a fetus

When the tenth week of pregnancy begins (fifty-six days after fertilization), the embryonic stage is over. The heart has been beating for a month, and the muscles of the torso, arms, and legs have begun to exercise. All the organs are in place, although they are still small and immature and far from coordinated in their functions. The embryo, now referred to as the fetus, has passed its first test with flying colors and will go on developing until it is ready to be born.

◀ **The brain**

The number of brain cells increases rapidly, and more and more cells interact.

◀ **The umbilical cord**

In this essential channel between mother and fetus, there are three ducts (one artery and two veins), through which nutrients are transported to the fetus and waste products removed.

◀ **The feet**

Feet lag behind hands developmentally, but now even the toes are beginning to take shape (see pages 142–143).

**The amniotic sac**

The fetus floats in a fluid-filled sac, the inner fetal membrane. An outer, somewhat more durable membrane also safeguards the fetus.

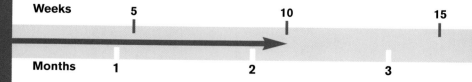

| Weeks | 5 | | 10 | | 15 |
|---|---|---|---|---|---|
| Months | 1 | 2 | | 3 | |

## Approximate size

Length 2.3–3 cm./0.9–1.2 in.
crown-rump length (CRL)
Weight 10–15 g./0.3–0.5 oz.

### Classic indications of pregnancy

The uterus now feels distinctly
enlarged upon gynecological
examination and softer than the
uterus of a nonpregnant woman.

### A room of one's own

The fetus has made itself at home
inside the protective shelter of the
uterus. The placenta, which has spread
across specific sections of the uterine
lining, has taken over production of all
the hormones needed to keep the fetus
comfortable and developing properly.
The progesterone-producing function
of the ovaries is now concluded.

### Twins?

Sometimes an ultrasound scan is performed at this time.
If there is more than one fetus in the uterus, they will be
seen. Twins may share the placenta or have separate
ones. Most twins have individual amniotic sacs, but in
rare cases they may share the same sac as well. It is still
difficult at this point to distinguish the individual organs
with ultrasound, but the cardiac activity is a good
indication of viable life, as is an occasional erratic
movement on the part of the fetus.

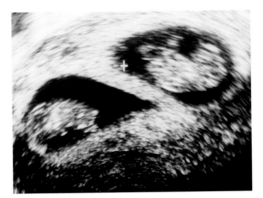

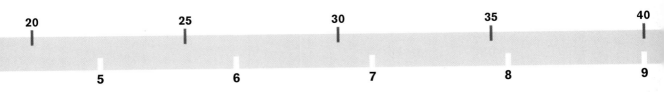

20    25    30    35    40

5    6    7    8    9

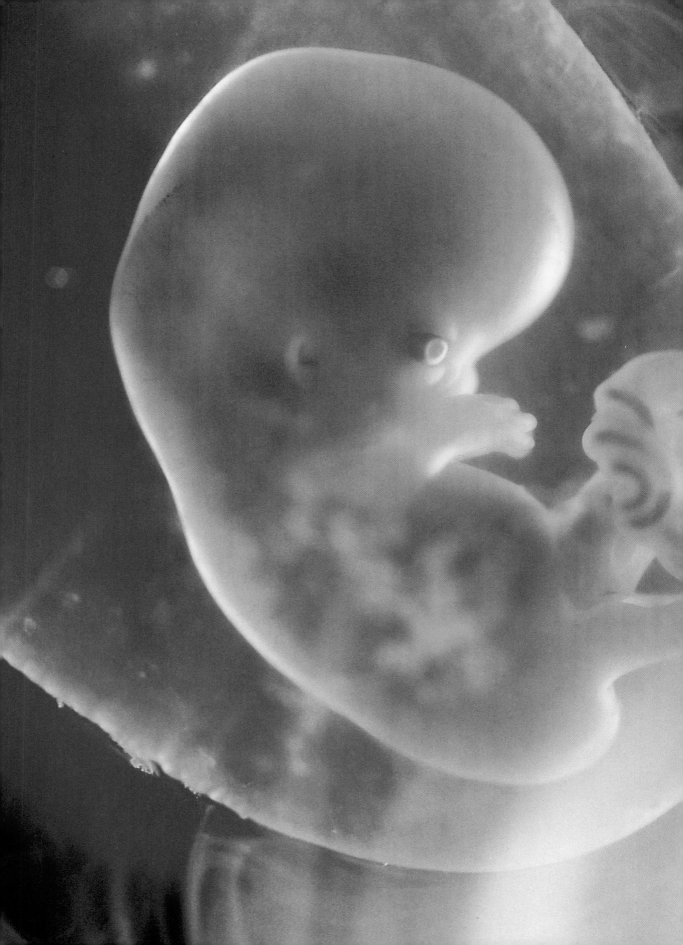

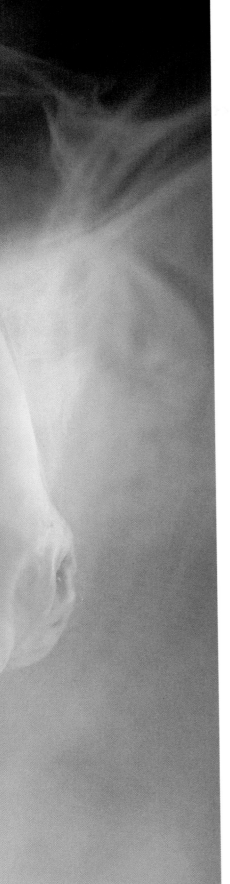

## Farewell to the embryonic stage

As the expectant mother enters the third month of pregnancy, the important task of establishing the fetus's organs has been successfully completed. For the fetus, now with very pronounced human features, the remaining task is to grow and refine its abilities, in preparation for the world outside the uterus. It still has several months left in which to continue developing in the protected environment of the womb.

**Back to the sea**

The origins of life are to be found in the sea— this newly created fetus swims in an inland sea with just the right temperature and salinity. Eight weeks have passed since conception, and the little hands are more like paws, with a suggestion of fingers.

## The first prenatal appointment

When ten to twelve weeks of pregnancy have passed, it is time for the couple's first prenatal appointment with a nurse and doctor (in some cases a midwife). The parents-to-be are usually very eager and curious, and the medical staff will also have questions for them: How have things been so far? How does the woman feel about her work situation? The answers often provide indirect information regarding how tired the woman has been and how much morning sickness she has experienced. Illnesses, prescription medications, and vaccinations are also important things for the medical staff to find out about for the record.

The largest number of questions from both sides will arise in a first pregnancy. When pregnancy is a new experience, the mother-to-be may have lots of questions and some concern about the delivery. If she is over thirty-five years of age—which is becoming more and more common in developed countries—she will undergo a more extensive examination. After that age there is a slightly elevated risk of chromosomal aberrations in the fetus, particularly Down syndrome. Today, however, many prenatal diagnostic tests can reveal genetic disorders (see pages 126–127).

The medical staff will also ask questions about any previous pregnancies. Has the woman had a miscarriage or abortion for any reason? If she has children already, how did her previous pregnancies and deliveries go? Were there any difficulties?

If the woman has had a previous cesarean section, the staff will need to have as many details as possible. If the cesarean was performed because the pelvis was considered too narrow in relation to the size of the baby, perhaps the baby will have to be delivered by cesarean this time, too; but if this child is not as large, it may be possible for the mother to have a vaginal delivery. If the cesarean was performed as emergency surgery—such as for sudden fetal distress—and the same problem is not expected to arise again, it is possible to plan for a vaginal delivery. But there is always a higher state of readiness for a cesarean if such a situation has already occurred. Based on statistical evidence, medical professionals tend to be cautious about trying a vaginal birth after a cesarean.

**In good hands**

Parents-to-be are usually excited about their first prenatal appointment. At last they can get answers to all their questions. Doctors, nurses, and midwives are accustomed to sharing in both concerns and excitement, and the appointment usually has a calming effect (opposite and above).

**Is the mother feeling all right?**

The health of fetus and mother are closely interrelated. For this reason the first prenatal appointment includes a number of tests. It is important to take the pregnant woman's blood pressure and blood count and be sure there is no protein or sugar in her urine (below and right).

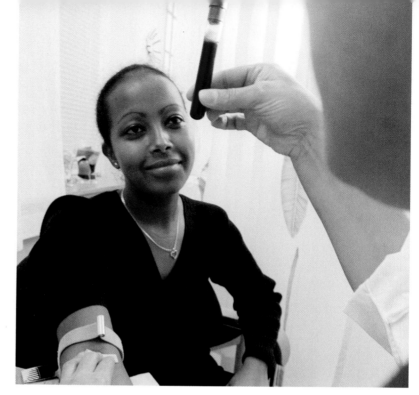

## Mother and child examined

At the first appointment a urine sample and blood tests will be taken. The urine sample will show whether any protein has leached out through the kidneys. If it has, the reason will be investigated. The blood tests will reveal the woman's blood group and whether she is immune to German measles, has had syphilis, or is positive for HIV or hepatitis. Gonorrhea and chlamydia tests are also done, particularly on young women. The consent of the woman is always obtained before testing.

Because the ability of the blood to transport oxygen is so important during pregnancy, it too is investigated by measuring the hemoglobin levels, or blood count. The blood count is followed regularly for the rest of the pregnancy. Hemoglobin production is contingent on access to iron, so it is very important that a pregnant woman gets enough iron, either through her diet or by taking supplementary iron pills throughout the pregnancy.

Blood pressure is monitored regularly during pregnancy. Women who have high blood pressure before becoming pregnant will need to continue treatment. High blood pressure that develops in the third trimester along with protein in the urine and generalized swelling (edema) is treated with medication and rest.

After all the tests have been taken, the woman is given a gyne-

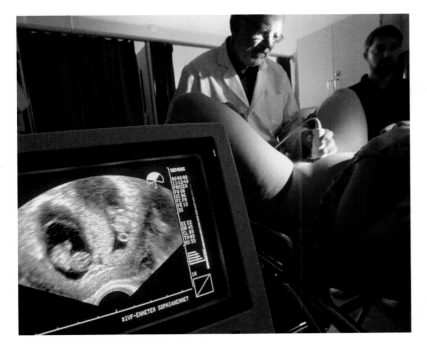

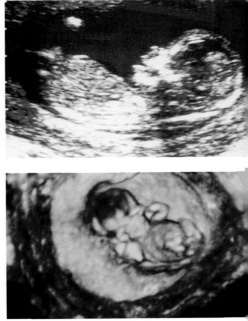

**The first ultrasound scan**

If there is any uncertainty about how long the woman has been pregnant, the first appointment is often concluded with an ultrasound scan (above). The ovaries will also be examined, to be sure there are no unwanted cysts in the woman's pelvic area.

cological examination. First the physician inserts a speculum into the vagina, to inspect the mucous membranes and the cervix, which protrudes down into the vagina. The cervix will be a little swollen, and the mucous membranes will be purplish in color. The doctor will look carefully for any possible pathological changes in the mucous membranes. If an infection is suspected, a bacterial swab of the vaginal secretion will be taken.

As part of the gynecological examination, the doctor will insert two fingers into the vagina toward the cervix, using his or her other hand to exert some pressure on the woman's abdomen, to feel the size of the uterus, and to check that the uterine lining is smooth, soft, and even. He or she will also check whether the ovaries feel normal.

The doctor will examine the woman's breasts carefully. A great deal of experience is required on the part of the physician to be able to distinguish normally enlarged mammary glands from any tumorous growths. If there is any uncertainty, the physician will arrange for an X-ray, or mammogram. A woman may have a mammogram at any time during her pregnancy with no risk to the fetus. Ultrasound scanning of the breasts is also becoming more common.

### Why won't she eat?

For many women, early pregnancy is a trying time. The man in her life may be enjoying his meals, while the expectant mother is often tired and nauseous. She pokes at her food and may even feel sick from the smell of cooking.

## Comfortable or miserable

During the first few months of pregnancy, most women's breasts are tender, and many women suffer from fatigue. The breasts may begin to change as early as a few days after ovulation. Many women have very little appetite and may even be nauseous, especially in the morning. These unpleasant sensations may begin after just a few weeks of pregnancy. The woman may feel she cannot possibly eat the big breakfast she usually enjoys, or that her coffee tastes peculiar. She may be oversensitive to all kinds of smells, particularly cooking odors. Many women become very sensitive to the smell of cigarette smoke.

Some women are so nauseous for the first two to three months that they lose rather than gain weight, both because they have difficulty getting anything down and because they tend to vomit up whatever they do manage to eat. Very occasionally a pregnant woman may need to be hospitalized for a few days to break a vicious circle. There is no really satisfactory explanation for the phenomenon of severe morning sickness. In fact, a woman may be extremely nauseous in one pregnancy and hardly at all in the next. The hormonal changes in her body are surely a significant factor, but anxiety and uncertainty in her new life situation are also aspects to be considered.

What should an expectant mother eat, and how much? One general guideline, for the entire course of a pregnancy, is to eat frequently and not too much at a time. Eating often keeps the blood sugar at an even level, which is an important way to avoid faintness. Eating a little at a time prevents stretching out the stomach, thereby avoiding putting pressure on nearby organs, which may be one cause of unpleasant sensations and nausea.

The issue of what foods are "dangerous" or unsuitable for a pregnant woman is much discussed. Although this kind of advice may often be useful, it is important not to be fanatical. The vast majority of women can go on eating approximately as they normally do, although it is always wise to heed warnings about additives that could be a health hazard either to the mother or to the fetus. Be cautious about eating raw or marinated foods that might contain bacteria. Unpasteurized milk products, such as mold-ripened cheeses, also belong in the less-appropriate-foods category. Avoid large predator ocean fish, such as swordfish, shark, or king mackerel (which may have large concentrations of mercury as well as freshwater fish, which may have high levels of PCBs or dioxin). Vegetarians and vegans may continue eating as usual but should be sure to get enough calcium and iron. Medical staff can provide advice about diet and any necessary supplementary vitamins and minerals.

Some pregnant women have almost irresistible urges for certain foods or sweets, such as licorice. Physicians interpret such desires as expressions of some dietary deficiency, such as iron or zinc; the cravings are for foods containing precisely the thing the pregnant woman needs.

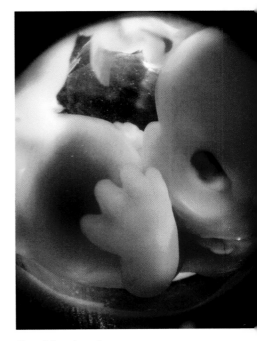

**Good food makes for positive development**

The most sensitive period is over, but the fetus will go on being affected by whatever the mother consumes. Prenatal appointments provide information as to what foodstuffs pregnant women should and should not eat.

**58 days, 3.4 cm./1.33 in.**

## The environment of the womb

Embryonic development follows a detailed pattern day by day. This pattern is determined by the genetic code, yet even before birth there is an intimate interplay between nature and nurture. An unfavorable uterine environment may mean an increased risk of illness later in life. For instance, inexplicably low birth weights in boy babies carried to term have been related to reduced sperm production and impaired fertility in adult men.

Although the placenta plays an important role in protecting the developing embryo and later the fetus, it is not infallible. Drinking, smoking, and certain medicines are known to have a negative impact on the uterine environment and thus on the baby-to-be; so too may maternal deficiencies of nutrients, vitamins, and minerals. The fetus that receives overdoses of sugar and fat may be overnourished, potentially causing damage that, if not visible at birth, may appear, more frequently than was previously thought, later in life.

**Out drinking–juice!**
Most pregnant women avoid alcoholic beverages and cigarette smoke, knowing they can be dangerous for the baby.

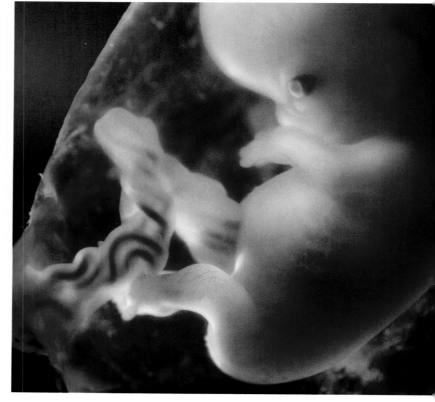

## Drinking and drugs

The safety of consuming alcoholic beverages in pregnancy has been much debated, and today the consensus is that heavy consumption of alcohol means a risk of serious fetal damage. For this reason most pregnant women refrain from drinking alcoholic beverages.

Children of women who have been heavy drinkers during pregnancy often have a particular appearance, such as severe squinting. But most fetal alcohol damage, known as fetal alcohol syndrome, is invisible, including irreversible brain damage. An occasional drink in early pregnancy, however, is not a reason to terminate a pregnancy.

Narcotic drugs, particularly heroin, are also known to have negative effects and to injure a fetus.

## Smoking

It is strongly recommended that pregnant women do not smoke. Women smokers run the risk of having growth-impaired children, and low birth weight is generally considered a risk factor for many illnesses later in life. A link between smoking and childhood asthma has been found, and recent research shows that this risk may arise even in the uterus.

## Medications and vaccinations

Some drugs are kept from reaching the fetus by the protective filter of the placenta; others pass through but are harmless. But a few medicines may both pass through the filter and be harmful to the fetus. The most vulnerable time is during the first eight weeks of pregnancy, when all the organs are being established. During this period it is important for a pregnant woman to consult a physician before taking any medication whatsoever, to be sure it is not harmful.

Live vaccines are unsuitable for pregnant women and should be avoided during pregnancy. Flu vaccines, however, are recommended.

## X-rays

If a woman of fertile age needs to have an X-ray of the stomach, intestines, gallbladder, kidneys, or lumbar vertebrae, the doctor will always ask whether she is pregnant, as such X-rays are known to be potentially harmful to the embryo. X-rays of arms and legs, mammograms, and dental X-rays, which do not expose the uterus to radiation, may be performed at any time during pregnancy.

## Stress

Stress in the woman's external environment is receiving increasing attention as a risk factor in pregnancy. Today stress is considered to influence both miscarriages and early labor.

## Illness and disease

German measles (rubella) is the best known infection that can harm an embryo. If a woman contracts rubella in early pregnancy, the baby's hearing may be impaired, among other things. The onset of symptoms like diarrhea, a high fever, and an itchy rash all over the body could indicate a disease with negative effects on the fetus. With any such illness it is important to contact the doctor for a reliable assessment.

## Hazardous substances

There is a great deal of concern today regarding how hazardous substances used in industry and agriculture can harm the uterine environment during pregnancy. Apart from well-known toxins such as DDT, PCBs, and radioactivity and certain neurotoxins like lead substances, links with fetal abnormalities have been difficult to prove; but this may be due to the fact that it takes a long time to collect such information, and that it is necessary to study large groups of children to obtain reliable statistics. Avoiding exposure to toxic substances, including pesticides, at work and at home, is a sensible precaution. Avoiding exposure to industrial solvents is very important, especially in the first trimester.

## Overweight

Recent research has shown that women who are considerably overweight when they become pregnant have a higher risk of fetal disorders. This risk is considered at least as great as the risk associated with smoking.

# expanding abdomen

**Suspended in a private space**

In the fourth month of pregnancy, the risk of miscarriage is substantially reduced. The fetus is resting comfortably in its own rosy universe, the yolk sac like a full moon overhead (opposite).

## The new fetus

Ten weeks have passed since conception, and twelve since the woman's last menstruation. The expectant mother is about to enter the fourth month of pregnancy, so the most vulnerable phase of the pregnancy is over: after the twelfth week miscarriage is relatively infrequent. In the uterus, surrounded by amniotic fluid, with a heart that beats twice as fast as that of its mother, a human-being-to-be who has just been through the first prenatal exam is napping. The ultrasound scan showed that the fetus could move, but it is still so small and its movements so tiny that the mother cannot yet feel its active presence.

The fetus is still growing fast, however, and the uterus keeps pace. By the end of the twelfth week it is the size of a man's fist, and just a month later it has grown to the size of a honeydew melon. The placenta too is becoming bigger and thicker, in step with the fetus's demands for oxygen and nutrition. All the organs that were created during the embryonic stage are developing and growing, and the proportions of the little fetus are becoming more and more like those of a newborn. The head is still disproportionately large.

The gelatinous body is becoming firmer, as cartilage is transformed into bone. Cartilage is softer than bone and more malleable. The skeleton is first made of cartilage, then is gradually transformed into bone according to very specific patterns, as long tubes become arms and legs. Calcification begins in the midsection of a bone, while the growth zones close to the ends of the bone remain soft and are not completely ossified until adolescence. Once cartilage becomes bone, lengthwise growth is halted. The extremely sensitive brain needs a protective shell, but it must be allowed to grow. In the fetal stage the skull consists of large, bulging bones that are loosely connected, so the brain has plenty of room to expand.

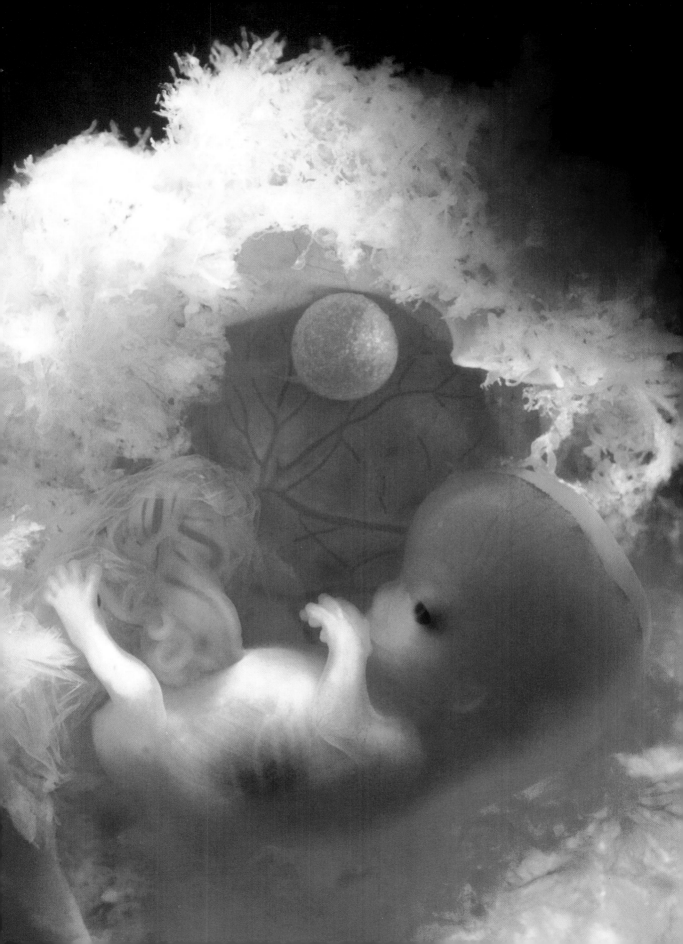

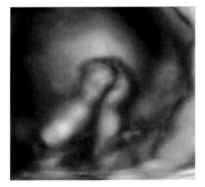

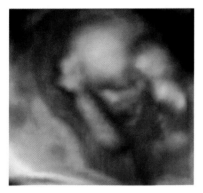

## Constant activity

The little fetus moves more and more every day, and the jerky body motions during the embryonic stage are now replaced by slower and apparently more goal-oriented movements. The hands often find their way to the mouth, ultrasound scanning shows, and the arms and legs are stretched and bent. The occasional breathing movement also appears; the fetus can be seen to yawn or hiccough now and then and seldom lies perfectly still for any length of time. The pattern of movement is much the same day or night; not until later in pregnancy will the fetus display a more regular day-and-night rhythm, with extended periods of sleep. Early on it never seems to sleep for more than a few minutes at a time.

Is it a girl or a boy? The external genitals have begun to grow (see pages 158–159), but the sex of the baby is still difficult to determine by ultrasound screening. In a boy fetus the testicles have already begun producing testosterone, and the primitive ovaries of a girl fetus already contain millions of immature eggs.

**The skeleton stabilizes**

The supple body becomes more stable as some of the cartilage begins to turn to bone. The calcification process begins in the long arm and leg bones, visible here through the thin skin (right and opposite). The fetus is now approximately thirteen weeks old and in constant motion.

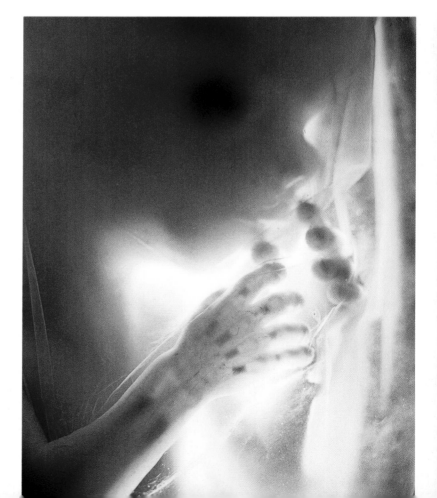

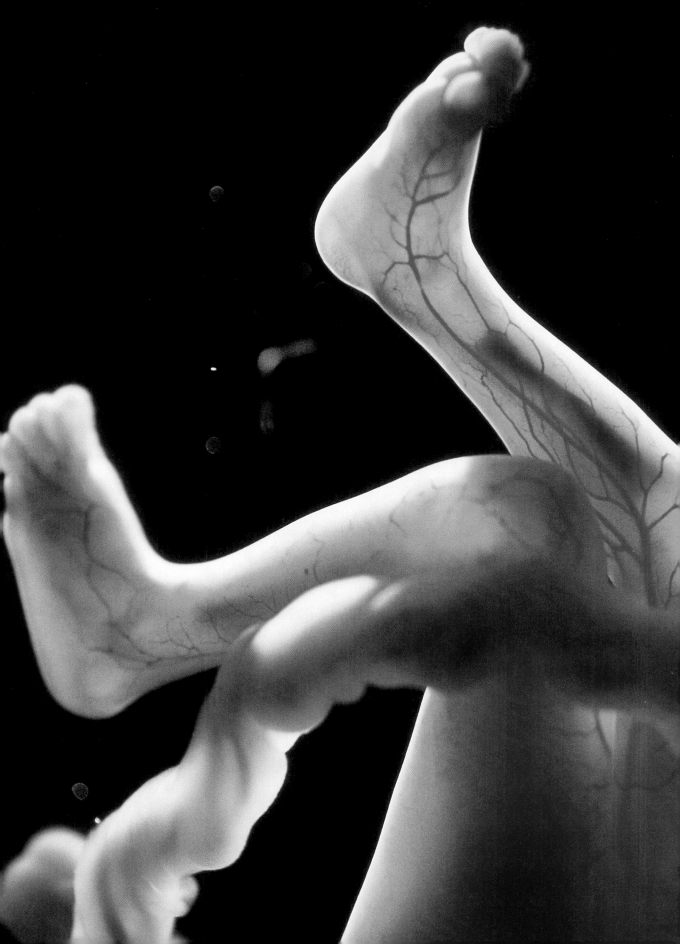

### Having a "picture taken"

Seeing one's unborn child appear on an ultrasound screen, a shadowy form with floundering arms and legs, is a major experience. This may be the moment—with photographic evidence in hand—when the expectant parents fully realize what is in store for them! From a medical point of view, an ultrasound scan is done to see if there is anything out of the ordinary. The physician studies the fetus carefully, explaining what is seen in the picture: the big, rounded head, the rapidly beating heart.

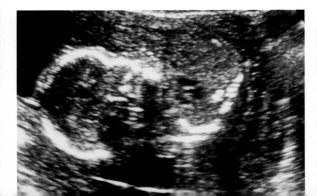

## Everything all right in there?

Today expectant parents often express overt concern, more frequently than in previous generations, that things might go wrong during a pregnancy and the baby might not be healthy. This is probably largely due to our increased knowledge about the causes of deformities and other disabilities.

An early miscarriage is nature's way of ensuring that an embryo with seriously damaged genetic material will not go on developing. But if the damage is less severe, the pregnancy will continue, and the parents may face a difficult decision. Three percent of all babies born have some little aberration or deformity, such as a deformed outer ear or a disfiguring birthmark, which may upset the parents and affect the child's appearance. In about one percent of cases, however, the deformity is more severe, such as a heart defect, or an abnormality of the abdominal or intestinal system or of the urinary tract. In such cases surgery will be necessary. Spina bifida is a deformity that often requires repeated operations and long periods of hospitalization and produces a great strain on both child and parents.

The most common genetic disorder is the one that causes Down syndrome, once known as mongolism. Children with this syndrome have an extra chromosome in pair 21, and thus have forty-seven instead of forty-six chromosomes. Information from the extra chromosome influences development, resulting in superficial external anomalies (slanted eyes, deformed ears, and abnormal hands) and many internal ones, affecting intellectual ability among other things. The risk of having a child with Down syndrome is greater in older women, but it occurs even with young couples. Because the majority of all babies are born to younger parents, most parents of children with Down syndrome are young enough to be in an age group that, in most countries, does not routinely receive prenatal diagnostic testing. In any case, not all couples who receive this diagnosis decide to terminate the pregnancy. Parents and siblings of children with Down syndrome have often testified about how these children enrich the lives of the other family members. Moreover, the degree of mental disability associated with the syndrome varies widely.

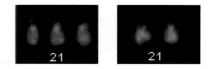

**One chromosome too many**

A single extra chromosome in pair 21 is the only genetic aberration in Down syndrome, yet the result is an individual very different from the norm. In recent years many countries have begun to offer all pregnant women early prenatal diagnostic testing, to locate cases where there is reason to suspect Down syndrome. For example in Denmark a blood sample is taken and studied for possible increased risks, and in Sweden and England ultrasound scanning is performed to have a close look at the neck of the fetus in week 12 or 13. Such scans reveal most cases of Down syndrome and other genetic disorders, often eliminating the need for amniocentesis.

## Better able to monitor developments

Rapid advances in genetic engineering have provided more and quicker ways of determining that all is well with the developing baby. Using amniocentesis, the structure of the chromosomes can be examined in detail. Even minuscule aberrations can be discovered, by using specific fluorescent techniques.

## *Prenatal diagnostics*

Today many types of prenatal diagnostic tests make it possible to determine whether the developing child has a congenital disorder. The most common is ultrasound screening, of which many women have several during their pregnancies. Amniocentesis has become more common now that many women are giving birth later in life. But the new opportunities to discover even minor aberrations may pose problems for physicians. Although prenatal tests sometimes lead to definite conclusions, at other times it is impossible to know precisely how a genetic aberration found in utero will affect the child when it is born.

If testing indicates that a fetus has a serious genetic disorder, the woman is usually advised to terminate the pregnancy. Most pregnant women who have prenatal tests, particularly ultrasound, consider the testing to be a way of obtaining extra reassurance that all is well and have not usually worked through how they might react if some disorder were actually found. For this reason, physicians should always discuss the possible consequences of such examinations thoroughly with the expectant parents beforehand. Some kinds of testing may imply a risk of miscarriage, and the couple must be informed of that possibility.

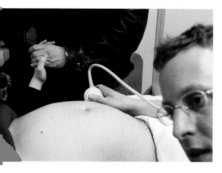

## Ultrasound scanning

Many countries require that every expectant mother be offered at least one ultrasound scan, including organ screening, during her pregnancy. In practice, though, ultrasound has come to be used much more frequently, now that it is known to be relatively harmless. Since occasional reports still suggest possible small risks, it is wise to exercise moderation and perform tests only when they are medically justified.

A state-of-the-art ultrasound monitor can show the head, chest, beating heart, torso, arms, and legs as early as week 11. The fingers can be counted and the genitals seen. In most cases it is possible to see if the baby is going to be a girl or a boy in week 17.

Ultrasound screening gives the physician the opportunity to look for things the untrained eye cannot see. If there is any indication of a serious problem, the examination may be repeated after a few days, possibly at a hospital with more advanced equipment. Or an amniocentesis may be used to check for chromosomal abnormalities; a fetoscopy (see page 128) will give an even more detailed examination. On rare occasions the expectant parents will be given news of a severe, incurable disorder. Depending on their personal beliefs and the laws in their country, they will then have to consider whether to terminate the pregnancy.

During the early months of pregnancy ultrasound examinations are usually performed vaginally, but as the fetus grows an external examination, performed by running the transducer over the mother's abdomen, becomes more informative. Sometimes both kinds of examination will be performed, depending on what information is needed.

## Chorionic villus sampling (CVS)

Another method that has become more common in recent years is chorionic villus sampling. CVS passes a thin needle into the placenta, guided by ultrasound, to get a sample of tissue called the chorion. Villi are tiny fingers that exchange oxygen and waste between mother and embryo. By sampling them—their genetic makeup is identical to that of the embryo—it is possible to determine the genetic code of the baby-to-be. CVS is usually performed during the tenth or eleventh week. In experienced medical centers the risk of complications, including miscarriage, is about the same as for amniocentesis.

## Amniocentesis

Amniocentesis is to date the most frequently used test to investigate possible fetal genetic disorders. Many women over the age of thirty-five are offered an amniocentesis, and more expectant mothers, irrespective of age, may soon be given this option.

The test is carried out in weeks 14 to 18. Cells shed by the fetus are found in the amniotic fluid; by extracting 10–20 milliliters (3–6 ounces) of this fluid, using a hollow needle and a fine syringe, the fetal cells can then be cultured in the laboratory for a couple of weeks, after which it is possible to determine their chromosomal makeup. Each pair of chromosomes is then examined. Amniocentesis can indicate whether the fetus is a boy or a girl.

The long waiting period for test results can be a terrible strain on the parents-to-be. New genetic engineering technologies make it possible to find the answer to a particular, well-defined question within just a few hours, without giving an overall picture.

## Triple screen test

This test measures a certain protein and hormones in the mother's blood to estimate the risk of Down syndrome, spina bifida, and another chromosome defect called trisomy. It is done between weeks 16 and 18.

**An excess of amniotic fluid**

This pregnant woman's body was producing far too much amniotic fluid, posing a danger to her own health and to that of the fetus. After a twenty-minute fetoscopy, the doctors were able to give her the go-ahead to continue the pregnancy—with some of the fluid drained off daily. Thanks to close supervision and a clear diagnosis, she was able to carry her baby successfully.

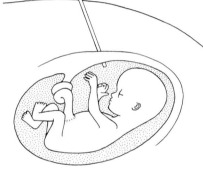

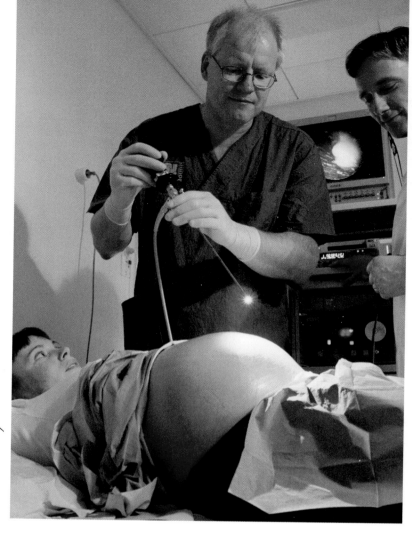

## *Focusing on the baby in utero*

If an ultrasound scan suggests the presence of a small, difficult-to-interpret defect, the expectant mother may choose to have a fetoscopy at a medical center equipped to do such a test. Using a very thin endoscopic instrument—the finest ones are less than 1 millimeter (0.039 inches) across, as thin as the average hypodermic needle—a doctor can examine the embryo or fetus up close in utero and detect very small defects or deformities. In the future doctors may even be able to perform endoscopic microsurgery on these tiny patients, in order to prevent miscarriages or keep a minor defect from developing into a major or even incurable one. Fetal surgery is already being performed at the most advanced hospitals in the world, for example on fetuses found to have spina bifida.

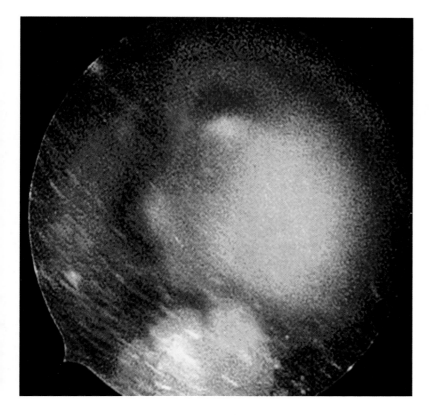

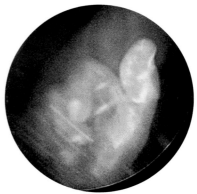

### A peek in the fetoscope

Tiny flakes of tissue drift past like snowflakes. In the cloudy fluid the facial features of the baby can be clearly discerned, a hand pressed against the nose, a tiny fist with the thumb standing out. Four months later this mother—whose body was producing too much amniotic fluid—gave birth to her baby. A perfect little boy!

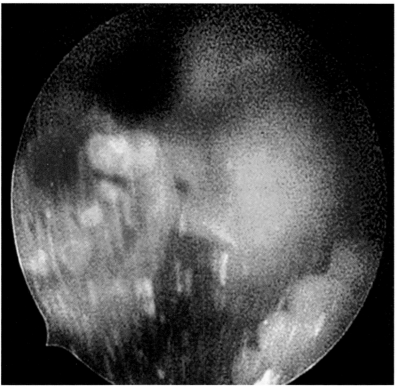

## An aqueous life

The fetus is encapsulated in the liquid-filled amniotic sac. Evidently human beings are constructed to spend the first nine months of our lives in water, just as all animals, at the dawn of evolution, lived in the original sea and only gradually became adapted to life on dry land.

Early in pregnancy the amount of fluid in the sac is negligible, but it increases with every passing day, to allow more freedom of movement for the growing fetus. The liquid is not crystal clear but is full of suspended flakes. The fetus swallows amniotic fluid, which passes through the gastrointestinal system, resulting in waste products. Other flakes are actually little clumps of cells from the lungs—a result of the fetus having swallowed fluid into the pulmonary system and then coughed it up. In addition, superficial cells flake off the skin of the fetus and float in the amniotic fluid. These cells can be used for prenatal diagnostic testing, as described above. The enclosed sea also contains the urine of the fetus, filtered by its kidneys. In spite of all this "pollution," the amniotic fluid is perfectly sterile and is completely renewed every five to six hours.

The skin of the fetus is well adapted to an aqueous life, protected by a waxy substance called vernix. When the skin is more or less fully developed, the very first hairs take shape, as early as the twelfth week of pregnancy. The first hair is known as lanugo, or "woolly hair," and covers the whole body, with a characteristic pattern on the skin. Around weeks 16 to 20, the roots of the hair on the head and in the eyebrows become a little thicker and more distinct. Pigment cells also begin to color individual strands of hair.

The lanugo falls away before the baby is born. Its precise role during the time in utero has not yet been determined, but it may function as a sort of binder for the protective vernix, which is produced in large quantities from the sebaceous glands, located along the roots of the hair. This sticky salve also protects the fetus against rubbing and against small superficial skin sores. Some of the vernix is rinsed off the skin and mixes with the amniotic fluid. The vernix covers the fetus throughout pregnancy, so the baby is generally both slippery and a little sticky at birth.

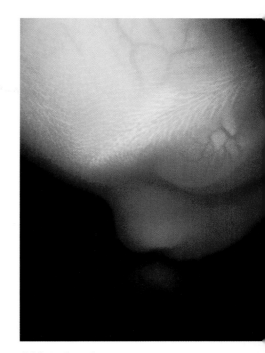

**A hint of eyebrows**

The ingenious fetoscope enables us to see the face, and even the tiny hairs on the eyebrows, of a fetus sixteen weeks old (above). The roots of the hair are already organized in a special pattern, although there is as yet very little pigmentation.

**Week 17**

The creation process continues unobtrusively (opposite). The fetus now measures some 18 cm./7 in. from head to foot, and facial features are beginning to show.

## week 17

# Nearly halfway

The complex process of creating a human being is now well under way. There is still plenty of room in the uterus for the fetus to move freely and shift position. It is virtually in perpetual motion, kicking at the uterine lining. The expectant mother will begin to feel these kicks around weeks 16 to 20.

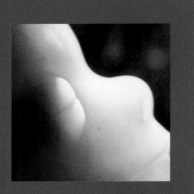

◀ **The eyes**

The eyelids are now complete, covering the eyes, which will not reopen until week 26.

◀ **The hair**

There are already tiny hairs on the face, and the entire body is covered with the fine, downy layer of lanugo hair.

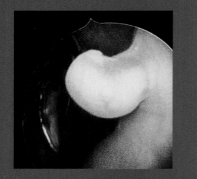

◀ **A boy**

The fetoscope clearly reveals the genitals of a fetus in week 17 (see pages 158–159).

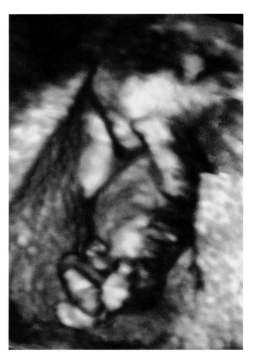

### Answers from ultrasound

In an ultrasound scan at around week 17, careful measurements are made of the length of the thighs, as well as the circumference of the torso and the head. All deviations from the norm are registered and painstakingly analyzed.

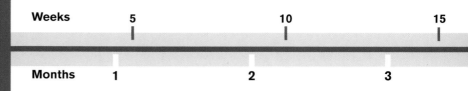

| Weeks | 5 | 10 | 15 |
|---|---|---|---|
| Months | 1 | 2 | 3 |

## Approximate size

Length approx. 18 cm./7 in. (Beginning around week 14 or 15, the length of the fetus is measured from head to foot.)
Weight approx. 90–120 g./ 0.2–0.26 lb.

## Fine limbed

The body is slender, and the arms and legs spindly. There is virtually no subcutaneous fat, so the skin is slightly wrinkled. The arms are now long enough to reach each other.

## Head circumference

The fetus's large head accounts for about a third of its entire length. Loosely joined skull bones protect the brain. The large soft gaps known as fontanels remain until the baby is about two years old.

## Nutrient depot

The placenta grows in pace with the fetus, occupying more and more of the space in the uterus.

## Steadier legs

The calcified leg bones are visible through the thin skin. The rest of the skeleton is still made of soft, pliant cartilage.

## The uterus swells

The thick muscles of the uterine wall have been stretching, and the uterus is beginning to compete for space with the woman's other internal organs.

| 20 | 25 | 30 | 35 | 40 |
| --- | --- | --- | --- | --- |
| 5 | 6 | 7 | 8 | 9 |

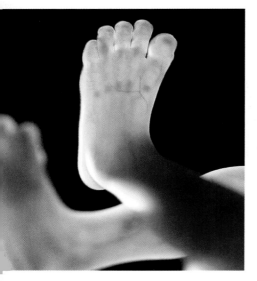

**The first kicks**

There is something magical about the first time an expectant mother feels her baby kicking. Suddenly this invisible family member is a physical presence. The early movements, which feel like fluttering butterfly wings, soon develop into stronger thrusts. Late in pregnancy the baby's kicks may be so hard, they take the mother's breath away or wake her from sleep.

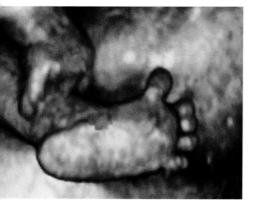

## Starting to show

In midpregnancy many women feel particularly well and find themselves in harmonious balance. Women who have had a great deal of nausea often feel better now, and fatigue declines. A pregnant woman's hair, eyes, and skin may appear very shiny and healthy, the size of her abdomen is not yet restricting her movements very much, and her respiratory system is not much affected, as plenty of room is still available in her abdominal cavity. She is sleeping well at night, although she may feel hot. The fetus may be compared with a furnace that is running full blast, day and night.

During the middle trimester of pregnancy the expectant mother undergoes visible changes, although she likely does not gain very much weight. Many factors contribute to determining how pregnant a woman looks at this point, including her height and her physical condition. In women expecting for the first time, the pregnancy usually begins to show later. The growth of the abdomen in pregnancy also reflects development of the fetus, which is why at prenatal visits medical staff regularly measure the distance between the upper edge of the pelvic bone and the upper edge of the rounding of the woman's abdomen (known as the symphysis-fundus measurement). This distance differs from woman to woman and from pregnancy to pregnancy, but the standard measure for week 20 of pregnancy is 18–20 centimeters (7–7.8 inches). In a first pregnancy, the distance increases by about 1 centimeter (0.39 inch) per week until the delivery.

Although the expectant mother is aware that her abdomen is growing and the ultrasound scan has shown her a little fetus with a heart beating very, very fast, until now she has experienced the fetus as lying completely still. Then one day she senses a soft fluttering in her belly. These first tangible movements may feel like gas bubbles rising and only gradually like definite bumps and kicks. The first signs are nothing a woman can be taught to feel; they have to be experienced to be known. A woman who has been pregnant before will recognize them more easily. As a rule, a woman who has had babies before feels the movements for the first time around week 16 or 17, while a woman expecting her first child is aware of them a couple of weeks later.

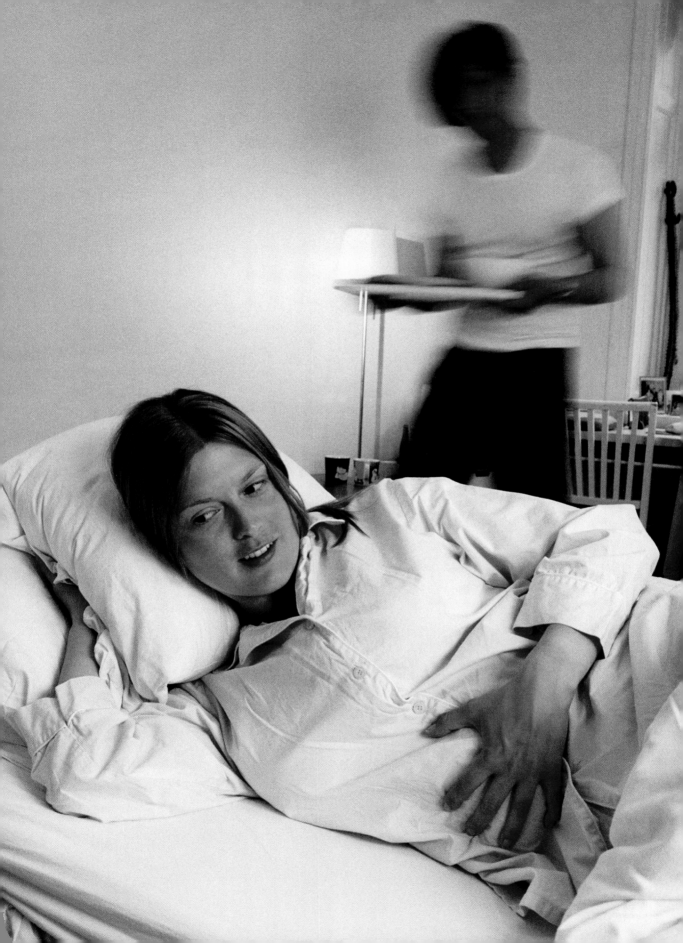

## A physical challenge

The middle trimester of pregnancy is a positive time for most expectant mothers, but minor problems do often occur as well. Low energy levels and tiredness may result from iron deficiency, as pregnancy consumes a great deal of a woman's iron reserves. Her own blood supply and that of the fetus are increasing, and iron is needed for this buildup. For this reason supplementary iron is often recommended early in pregnancy, but in reasonable amounts, as high doses of iron tend to cause constipation, which may be a problem in its own right.

Unpleasant vaginal discharges are common during this part of pregnancy. Sometimes they are caused by harmless fungal infections; from a medical viewpoint they are seldom cause for concern. Most fungal infections can easily be cured with medication that does not enter the bloodstream of the expectant mother and cannot harm the fetus. Occasionally discharges arise from a bacterial infection; these must be treated because they may have a negative impact on the cervix and cause early onset of labor.

In this phase of pregnancy some women feel short of breath. Although this sensation becomes more frequent later in pregnancy, a sudden exertion may bring it on now.

In the first three months of pregnancy, the woman likely gains only a few pounds, but soon she must keep an eye on weight gain, as it is easy to gain too much too fast. But she should not be so concerned about gaining weight that she deprives herself and the baby of the necessary nutrition. A weight gain of about 13 kilograms (28.5 pounds) at delivery is regarded as ideal. About half this added weight is accounted for by the baby, the placenta, and the amniotic fluid. Many women gain much more, and a few a little less. In early pregnancy a woman may lose weight due to nausea. In the last few months many women gain a lot of weight because they are retaining water. This happens because the circulatory and lymphatic systems lose their ability to manage water properly, but the normal balance is regained relatively soon after delivery (see page 163).

A pregnant woman should bring up any problems or concerns at her prenatal visits, when her blood pressure and her blood count are taken, her weight is checked, and if necessary, protein or sugar in her urine is tested.

**Approximate weight gain by the end of pregnancy attributable to:**

| | |
|---|---|
| Baby | 3.4 kg./7.5 lbs. |
| Placenta | 0.7 kg./1.5 lbs. |
| Amniotic fluid | 1.0 kg./2.2 lbs. |
| Uterine lining | 1.0 kg./2.2 lbs. |
| Breast size | 1.0 kg./2.2 lbs. |
| Mother's circulatory system | 1.4 kg./3 lbs. |
| Increased fluids in mother's tissue | 1.5 kg./3.3 lbs. |
| Mother's lipid depots | 3.0 kg./6.6 lbs. |
| Total | 13.0 kg./28.6 lbs. |

**Relax**

During pregnancy life is often topsy-turvy. There is no need for a pregnant woman to stop living an active life. Many expectant mothers jog, ride their bikes, or exercise in some other way far into pregnancy. Even if not all women can accomplish yoga poses as sophisticated as this one (opposite), yoga can be a very relaxing form of exercise.

## In good shape for childbirth

Keeping in good physical shape during pregnancy is very important, and if a woman has a sedentary job, it is especially important to get frequent physical exercise. Ordinary brisk walking is a good start. A woman with a regular fitness routine can usually continue it in the first part of her pregnancy. As she grows, however, she will find it necessary to alter her exercise patterns and perhaps even the kind of exercise she chooses.

Special prenatal exercise classes are available, in which the movements are specially adapted for pregnant women, including prenatal yoga and gym. Some of these exercises focus on getting the muscle groups that will be needed during the delivery into tiptop condition. A woman can work up the muscles of her pelvic

**Go with the flow**

In water even the clumsiest body becomes a lithe craft. Swimming may be the best form of exercise for a pregnant woman. In a heated pool she can imagine how her baby feels in the womb: enveloped, secure, and virtually weightless.

floor by learning to tighten them. At least during a first pregnancy she will probably need professional guidance to learn the right ways to do this. It is important to work with these muscles regularly both before and after delivery, so that the muscles that have been under so much strain can regain their original resilience. Some slight "stress incontinence" is quite common after childbirth, especially when coughing or laughing; these exercises can help solve that problem as well.

Many pregnant women complain of backaches. Pregnancy is a great strain on the back, as increasing belly size has to be counterbalanced by backward-leaning physical posture. Many good exercises can counteract back strain and should be started early in pregnancy.

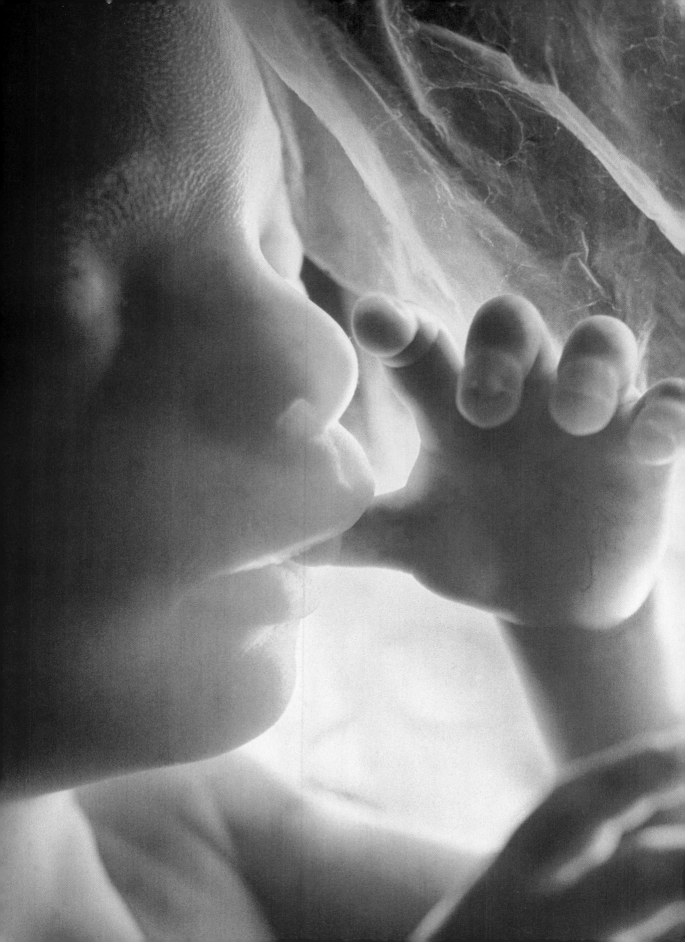

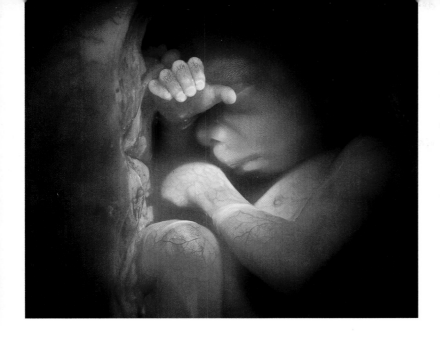

**Quiet but not silent**

Although the fetus is enclosed in the amniotic sac inside the uterine walls, it does not live in a world of silence. Still, not until weeks 18 to 20 (left) has its hearing developed sufficiently to take in sounds from outside the mother's body.

## Life in the womb

By midpregnancy the fetus has begun to explore its own body and environment using its hands. It often holds on to the umbilical cord, and when a thumb approaches its mouth, it will turn and begin to make sucking motions with its lips—a survival reflex. The baby must be able to grip and suck immediately after birth, and pull itself up to the mother's breast. So the fetus is in constant practice, kicking its legs and waving its arms. With every passing day the fetus becomes stronger and more agile.

The fetus is also using its sense of hearing for orientation. Its most familiar sounds are surely the noises of the mother's digestive system and the swishing from her major blood vessels, but gradually the fetus also begins to perceive the sounds of the mother's world, such as music and the father's voice. The eyes of the fetus are sensitive to light, even though the eyelids are still shut tight.

While even early in development the fetus swallows amniotic fluid, whether or how the fetus perceives taste and smell has been impossible to determine. We have no way of knowing whether the fetus tastes the slight salinity of the amniotic fluid. Still, we have indirect evidence that the fetus tastes and smells, since a newborn immediately reacts positively or negatively to tastes that are sweet, salty, or bitter, either on the mother's nipple or in breast milk.

**Thumb-sucking**

A fetus sucks its thumb in week 20, rocked to sleep in the protective chrysalis of the amniotic sac (opposite). Thumb-sucking is not a form of comfort, but a way of practicing a reflex in utero that will be important for survival after birth.

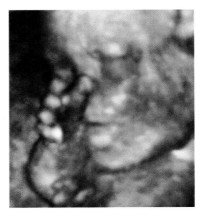

**Week 7**

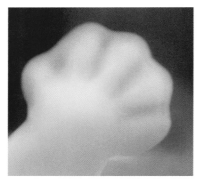

**Week 8**

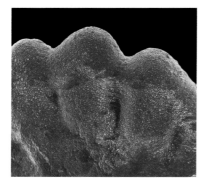

**Week 8**

## Hands develop before feet

By midpregnancy, the fetus's hands and feet are well developed, and the mother can feel the little knocks they make inside the womb when it moves. Hands and feet begin to take shape in the very early embryonic period; the hands develop a week or so prior to the feet, and the arms and legs grow out last, so that the proportions of the fetus gradually become more like those of a child. The fingers, which will eventually perform highly complex movements, develop before the toes. Although the toes play a subordinate role, they become very important at about the age of one, when children need to keep their balance and learn to walk.

As early as week 5, three weeks after fertilization, a clear line develops down each side of the body, from what will be the shoulders to what will be the hips. During week 7, two little skin-covered buds appear at either end of each line. These buds are raised but flattish, reminiscent of the mitts of a seal. Soon they develop an edge, which then begins to protrude. At the top end this protrusion signals the connective tissue to prepare the hands and upper and lower arms, and at the other end, somewhat later, the feet, thighs, and lower legs. By weeks 14 to 15 the hands are able to grip, although awkwardly, and the feet can kick.

The feet develop before the legs, as the hands precede the arms. The legs do not begin to grow until weeks 14 to 16 and remain malleable throughout the uterine period; they are often bent at sharp angles to give the fetus plenty of room to grow.

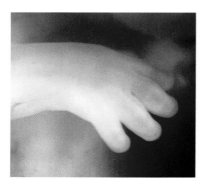

**Week 11**

### Hand development

A budding arm protrudes from the torso as early as week 6. Two weeks later there is a suggestion of fingers. The tissue between the fingers regresses. In the coming weeks the hand gradually takes shape, and in week 11 five little fingers can be counted on each hand. In week 19 the nails are also visible.

**Week 19**

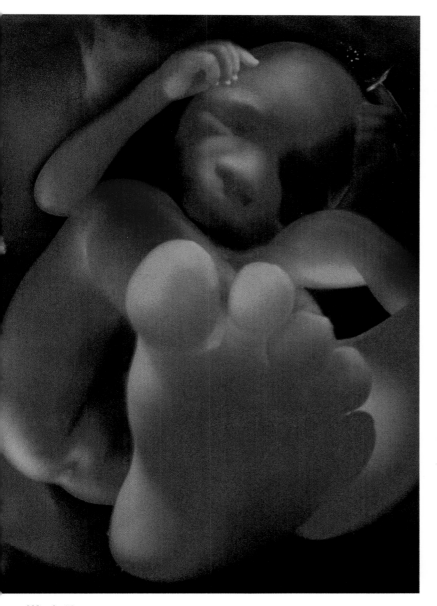

**Week 19**

**Week 6**

**Week 10**

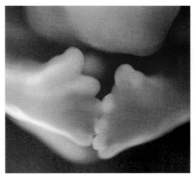

**Week 11**

### Foot development

In week 6 the primitive feet and legs are no more than buds protruding from the backbone. The vestigial tail vanishes early on, but we do retain a tailbone, a few small, stunted caudal vertebrae, all our lives. In week 11 the individual toes become slightly visible, though they are small and underdeveloped. The toes develop the same way as the fingers but slightly later.

**Week 14**

## A little world

Everything the fetus requires in the form of warmth, nutrition, and stimulation is available inside a round capsule of no more than 3–4 decimeters/10–15 inches in circumference. Although the fetus is beginning to resemble a newborn infant, the pregnancy is no more than halfway to completion, and the baby's weight is only 0.5 kg/just over a pound.

**Week 24, approx. 30 cm./11.8 in.**

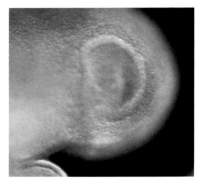

**Week 6**

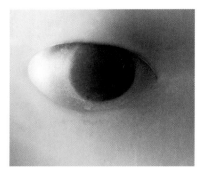

**Week 10**

**Week 12**

**Week 15**

## *The fetus can discern light ...*

Does a fetus see anything? It is known that the eye can sense light as early as the third month of pregnancy. Sometimes when an endoscope is inserted into the amniotic sac, a fetus tries to protect its eyes from the light on the instrument, either by turning away or by using its hands and fingers. Although the eyes remain shut until week 26, the thin eyelids do not close out light completely. Perhaps the fetus also experiences light if the pregnant mother lies on a sunny beach in a two-piece bathing suit, and she is slender. Most of the time, however, darkness prevails in the womb. Light impulses from the eye pass along relatively long nerve pathways to the vision center of the brain, which is at the very back, near the neck. As the brain and the retina develop and become coordinated, the vision center becomes able to classify impressions as light or dark, to sense nuances of color and shape, and gradually to structure complete, coherent visual impressions.

Life in the uterus does not require a highly developed sense of vision, yet the eyes begin to develop early in the embryonic stage. First the frontal portion of the brain sends out a hollow shoot toward the skin on either side of what is going to be the face. The end of this stalk is a puffed little bubble, the beginning of the eye. When this bubble reaches the inside of the skin, it curves inward, like a cup. The bottom of the cup becomes the bottom of the eye, and the cover of skin is transformed into the retina. In the cavity of the cup a lens gradually forms, as well as a cornea. In front of the lens the iris takes shape, growing from the edges toward the middle. Last of all, two flaps of skin fold down to become eyelids. The eye is complete.

### Protective eyelids

The eye is clearly established by week 6, and soon the lens develops, a spin-off of the skin. In week 10 there are primitive eyelids. These quickly cover the eyes, which will not open again until week 26.

## ... and hear its mother's heart beating

The development of the outer ear, from a shapeless little bud to a finely chiseled, perfectly shaped organ, takes several months, and is less significant than the development of the inner ear. Both developments begin early and are quite similar to those of the eye, in that contact is first established between the sensory organ itself and the brain, which is the interpreter of its signals. A bubble is tied off on either side of the rear portion of the brain. This bubble eventually becomes the inner ear, containing the organs of both hearing and balance. Somewhat later the outer ear, with the auditory canal and the outside of the eardrum, begins to take shape. The intermediate section, the middle ear, with its auditory bones (the hammer, the anvil, and the stirrup), begins to project inward from the throat. It then takes a long time for all the creases and folds of the outer ear to take their final shape. Minor aberrations in the appearance of the outer ear are usually of no consequence for a child's health, hearing, or development.

Although the ear is structured early, it cannot perceive sound until weeks 18 to 20. The womb is not a silent world, and when the fetus's hearing begins to work, it can hear the mother's digestive system gurgling, her heart beating, and her blood vessels streaming. Her voice imprints early into its consciousness. Other sounds probably play no role as yet, but toward the end of the pregnancy external sounds penetrate, and many fetuses react, for instance, to music. The fetus may, like the mother, have a favorite melody. Loud sounds may both stimulate the fetus and expose it to stress, quickening its pulse considerably.

### The ears are established

In week 5 two tiny, clearly delineated hollows appear just below what will be the head. As the ears are being established, they are extremely sensitive to disturbance. If the woman catches German measles (rubella) early in pregnancy, serious hearing impairment in the child may result. We can tell little about how well an ear functions from its external appearance. The middle and inner ear govern both hearing and balance.

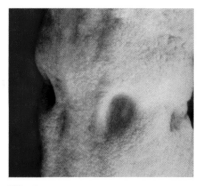

**Week 5**

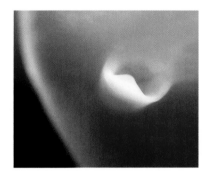

**Week 9**

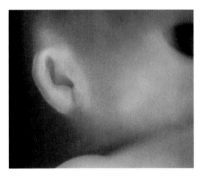

**Week 12**

**Week 18**

# Past the line

When the fetus has reached this point in its development, a major line of demarcation has been crossed. If, for any reason, contractions should begin now, and the fetus should be delivered, it has a chance of surviving if all the resources of the neonatal unit are made available. But most babies go on developing in utero. They grow quickly now, and the amount of free space declines, but they can still shift position quite freely.

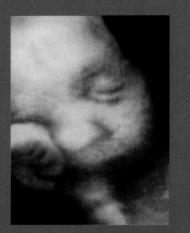

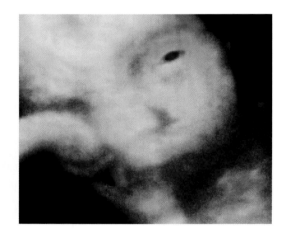

## A flexible cord

The blood vessels of the extremely supple umbilical cord are embedded in a gelatinous mass. This construction helps impede the development of knots that could otherwise result in a devastating inhibition of the blood supply.

## The eyes open

The eyelids remain closed and protect the eyes effectively until approximately week 26, when the nerve fibers to the eyelids allow the eyes to open and blink.

## ◀ Loosely wound

The long umbilical cord can sometimes twist around the body without impinging on the baby's freedom of movement.

## Room to maneuver

The baby can still change position in the uterus and busily practices various movements later displayed by the newborn. Lifting the arms and hands at the same time is part of a reflex, governed by the nerves. The legs are bent for reasons of space.

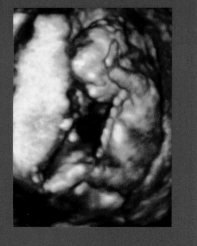

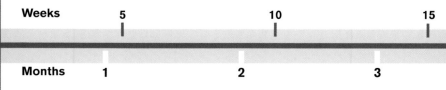

| Weeks | 5 | 10 | 15 |
|---|---|---|---|

| Months | 1 | 2 | 3 |
|---|---|---|---|

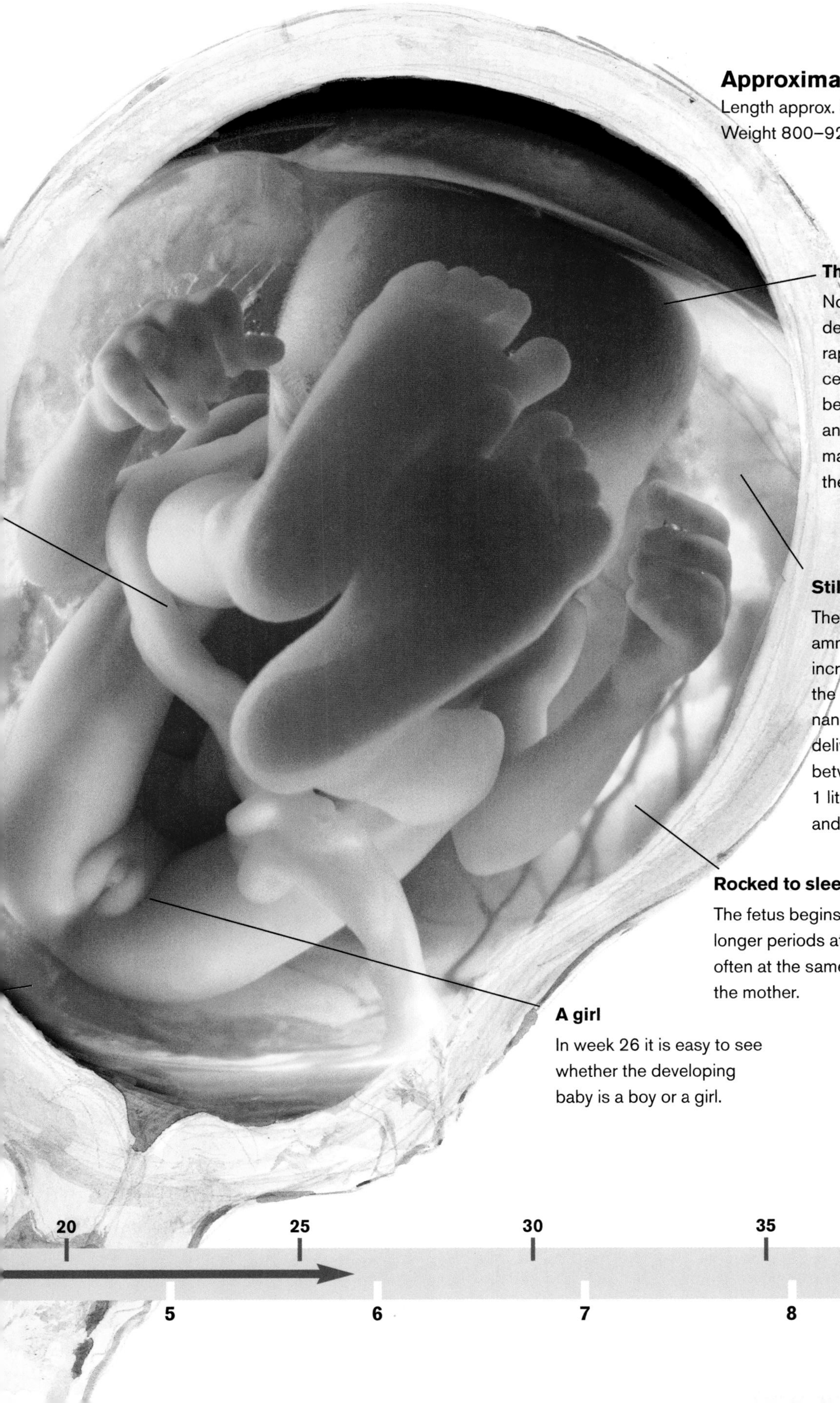

**Approximate size**
Length approx. 32 cm./12.6 in.
Weight 800–925 g./1.8–2 lbs.

**The brain**
Now the brain develops very rapidly. The cerebral cortex becomes furrowed and convoluted to make room for all the nerve cells.

**Still underwater**
The amount of amniotic fluid increases toward the end of pregnancy and at delivery is usually between 0.5 and 1 liter/about a pint and a quart.

**Rocked to sleep**
The fetus begins to sleep for longer periods at a time, often at the same time as the mother.

**A girl**
In week 26 it is easy to see whether the developing baby is a boy or a girl.

20   25   30   35   40

5   6   7   8   9

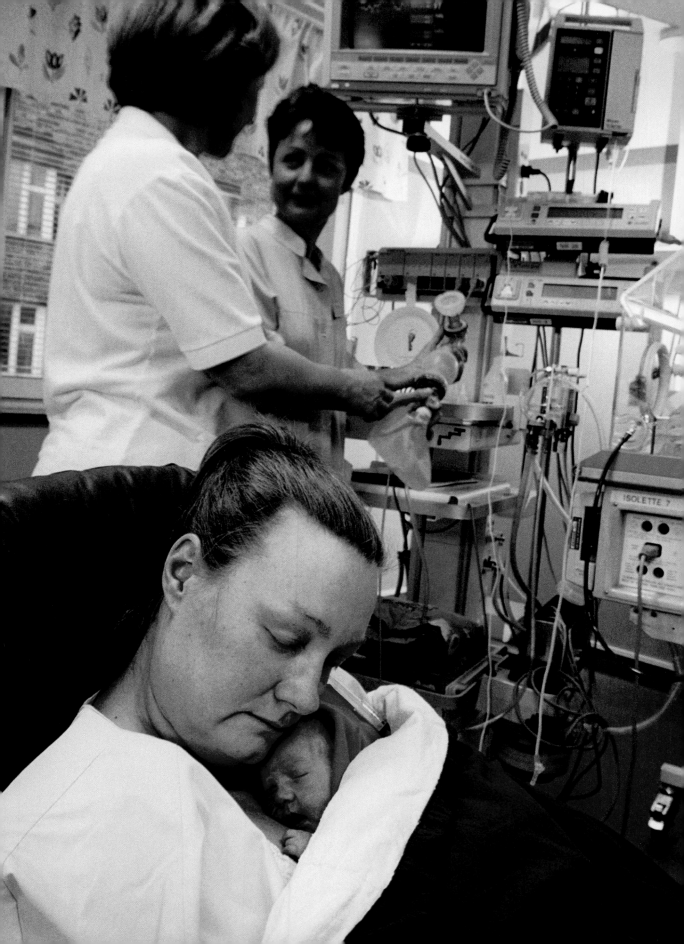

## Born early

Forty weeks after the first day of the last menstrual period is considered the length of an average pregnancy. But only slightly more than half this time in the uterus is absolutely needed if the fetus is to survive. Not many years ago a delivery prior to three months before the due date was a disaster. Today, at least in the developed countries, more than 90 percent of the babies born in week 28 survive. But the beginning of life—just like the end of life—is a difficult tightrope walk in terms of medical ethics. A premature baby may be able to survive, but only at the risk of grave vision or hearing impairments, or of severe mental retardation (if it proves impossible to protect the immature brain). How premature is too premature? Staff at intensive care units of neonatal wards must constantly consider and reconsider this question.

In recent years new understanding about the condition of the lungs in the fetal stage has made it possible to help even very small babies survive. Today there are children who have survived and developed normally who were born as early as week 23 and weighed less than 500 grams (1 pound). But immature lungs can cause irreversible brain damage owing to impaired oxygen supply. The lungs develop much later than, for example, the heart, which supplies oxygen from the placenta to all the organs in utero. Because the placenta transfers oxygen from the mother, and the oxygen supply travels to the fetus via the umbilical cord, the lungs serve no real function in the womb. Not until the moment of birth, when the oxygen supplied by the umbilical cord is abruptly terminated, are the lungs really needed.

The immature lung is subject to a condition called respiratory distress syndrome (RDS). RDS occurs when the lungs do not produce a lubricant (surfactant) needed for the air sacs to open and stay open, so as to transfer oxygen. The sacs also contain fluid. This is why a great deal of the life-saving effort that goes into helping babies born prematurely is focused on making the lungs sufficiently mature so that the baby will be able to breathe independently.

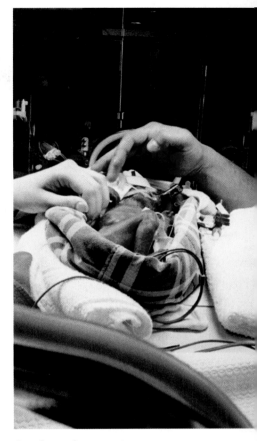

**A safe environment**

Gentle hands care for a premature baby. In a neonatal nursery premature babies receive the care they require from the professional staff and from their parents (opposite and above). The caregivers try to maintain an atmosphere of calm expertise, because it is particularly important for these babies to get off to a serene, secure start.

| Week | 23 | 24 | 25 | 26 | 27 | 28 |
|------|------|------|------|------|------|------|
|      | 44% | 63% | 73% | 81% | 88% | >90% |

**Survival rates**

for premature children from Swedish statistics for 1991–2000.

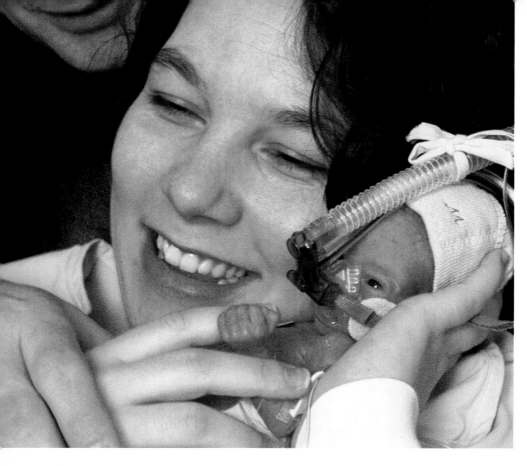

## Smilla—a survivor

Just a decade ago, Smilla would probably not have survived (left). She was born in week 26, weighing only 660 g./1.4 lb. She was kept alive and survived with no harm done, thanks to modern technology. Developmentally, she is expected gradually to catch up with babies carried to term.

## An artificial umbilical cord

The lungs of a very premature baby are underdeveloped, and so oxygen is delivered via a plastic tube through the nose and mouth. An electrode monitors the baby's heartbeat. These babies also need to be fed often but cannot be nursed because they are unable to suck properly. Instead, their mothers pump milk and use a spray bottle to feed their babies, an acquired skill. A premature baby needs warmth and security; the parents can provide both by cuddling the baby on their chests, a method known as kangaroo care (below, right and opposite).

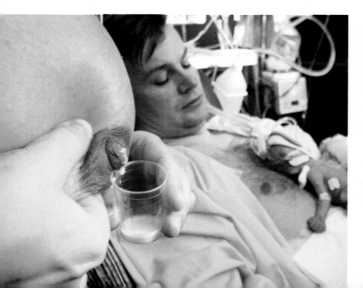

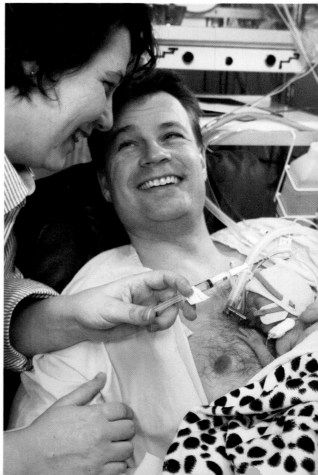

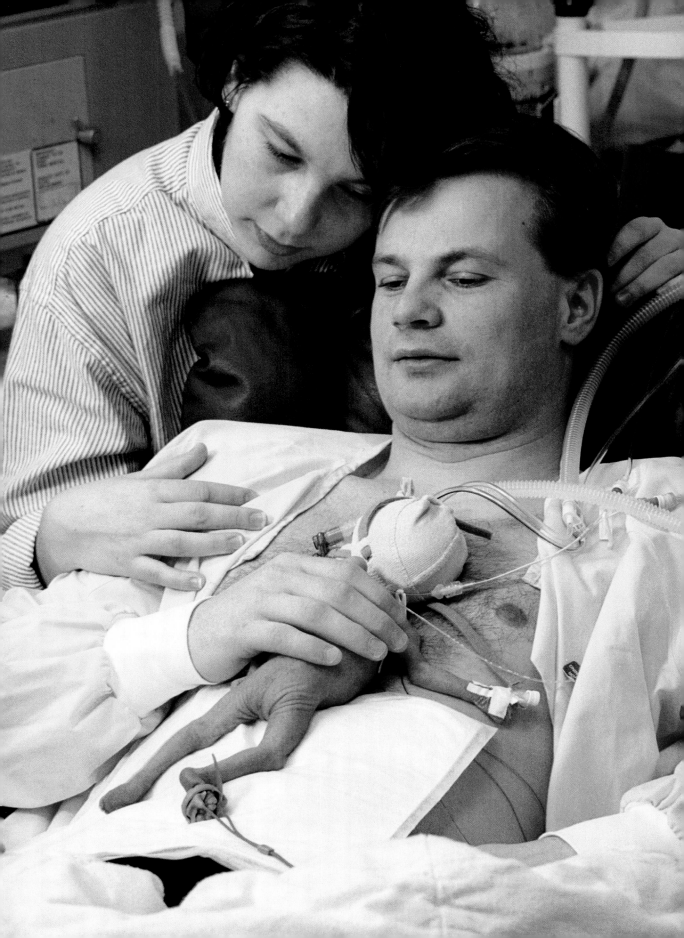

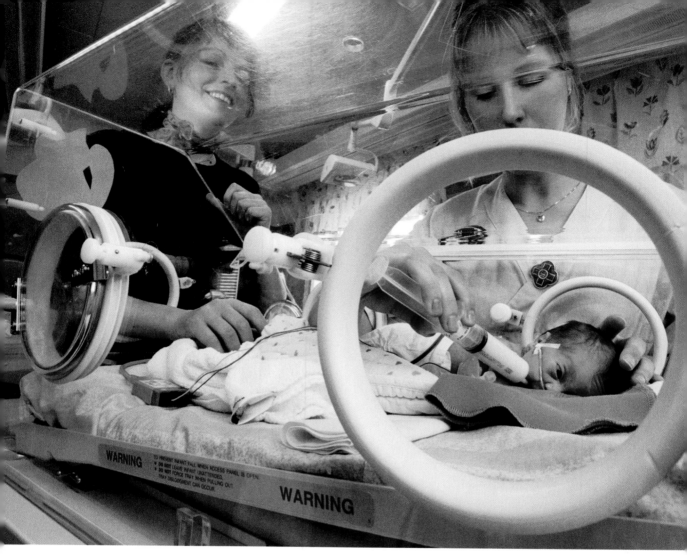

### An artificial womb

A state-of-the-art incubator provides a premature baby with an environment quite similar to conditions in utero (above and right). The incubator is snug, warm, and free from viruses and bacteria. It is much like a greenhouse. In an incubator babies often grow, thrive, and gain weight faster than they would have had they remained in utero.

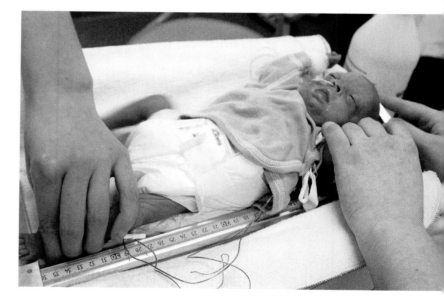

## Stronger every week

Today an obstetrician has several ways to prevent the delivery from beginning too early and to help an immature fetus remain longer in the womb. Bed rest for the mother is a method used for years, and effective drugs can counteract the tendency of the uterus to contract. When a woman has an uncommon condition called incompetent cervix, a procedure called cerclage is done. The cervix is sewn closed to provide mechanical support. It has to be done, of course, before the cervix is completely effaced during labor, and it is removed in the ninth month. As always in pregnancy, regular checkups are essential, particularly if there is any reason to anticipate the problem of early contractions.

Postponing delivery by as little as a couple of weeks may be just what is needed to allow the lungs to mature, making it easier for the baby to breathe and supply all the organs with oxygen. Treatments with the hormone cortisone speed up lung maturation. Even if the amniotic fluid has leaked out at an early stage of pregnancy through a hole in the amniotic sac, much can be done. Amniotic fluid is continually produced, and with bed rest and careful monitoring, labor may be postponed by several valuable weeks. But after rupture of the membranes, the risk of vaginal bacteria making their way into the uterus increases, sometimes necessitating antibiotics.

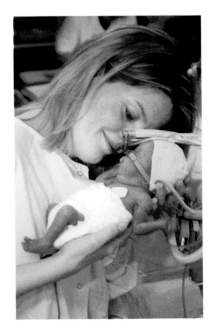

**Two but tiny**

Twins run a greater risk than other babies of being born early. Moreover, they have had to share both space and nutrients in the womb and often weigh at least 1 kilogram/ 2.2 lbs. less at birth. Both incubator and intimate contact with their mother are essential during this sensitive period (above and below).

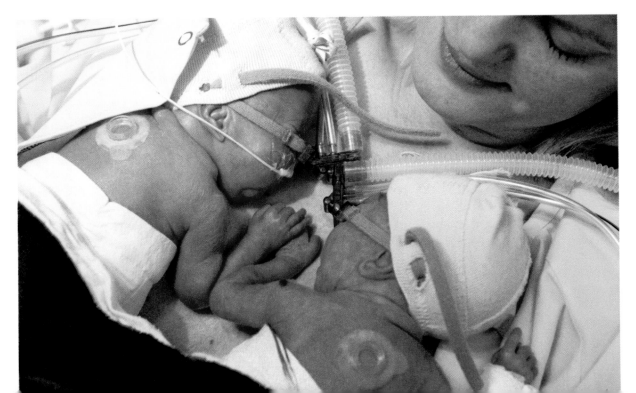

# Anticipation

**During the seventh month**

The baby's face is completely developed except that the cheeks are not yet filled out (opposite). This means that the eyes, which now open and shut, seem to protrude. During the remaining weeks in utero, the baby will gain a couple of kilos/about 4 pounds.

## Growth

Two-thirds of the pregnancy is now past, and for most babies life in the dusky womb with its long, pleasant health-inducing soak in the 37°C (98.6°F) amniotic fluid will continue for a few more months. Early in this period the uterus is still quite a roomy environment for the baby, but it will soon seem cramped, and many babies appear to prefer lying head downward. The baby can still turn around, but usually by week 36 it has decided whether to be born head or feet first.

The baby is still slim and has not yet accumulated fat under its skin, which is red and thin. Not until the last four weeks of pregnancy does the baby really fill out to fit its skin and develop that plump newborn appearance. Still, weight gain is continual, about 200 grams (7 ounces) per week. The organs, including the lungs, are maturing steadily, and breathing practice begins to play an important role. Sometimes breathing practice results in hiccoughs, experienced by the expectant mother as rhythmic abdominal twitches. When the baby's feet touch the uterine wall, its walking and crawling reflexes are triggered. The sucking reflex is a skill that requires constant practice because it will be needed right after birth.

When the mother-to-be enters week 27 of pregnancy, the baby's eyes and eyelids are so well developed that the eyes are now open and blinking at regular intervals. If the eyes open later, the fetus may have been exposed to something harmful during development. Such damage is most often alcohol-related. Hearing is developed and refined during this period, and the baby likely perceives more and more sounds from the outside world.

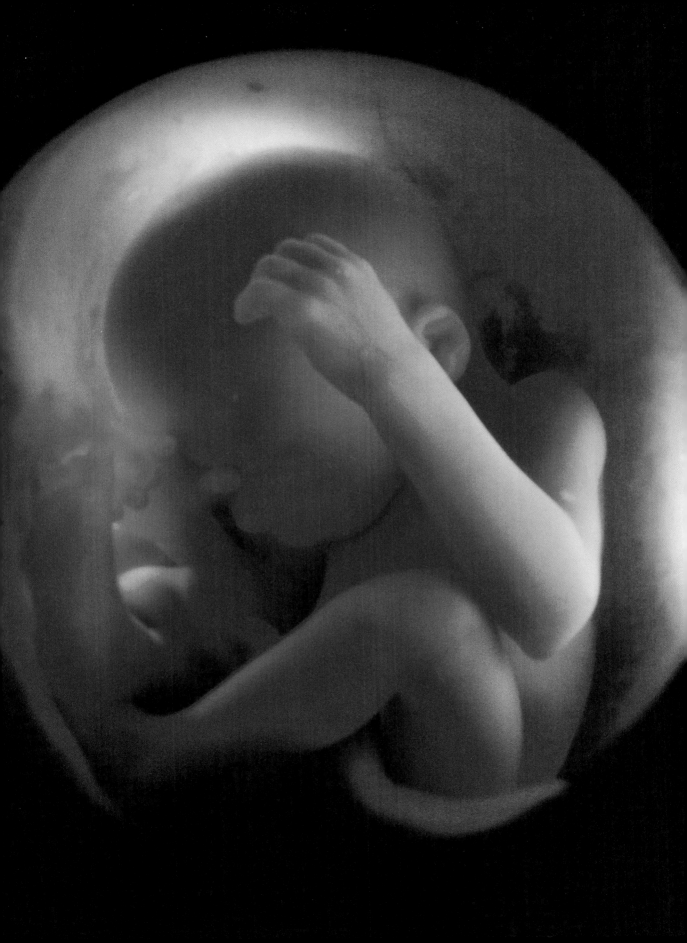

## Initially alike

In week 9 the genitals of boys and girls are very similar (right). The little bud with walls on either side will eventually develop into either a clitoris or a penis.

## A girl or a boy?

As described earlier, the genetic code determines the sex of the baby, the twenty-third pair of chromosomes in girls having two X chromosomes and in boys one X and one Y. A particular gene in the Y chromosome, known as SRY, is known to be essential for the development of the fetus into a normal male child. Another important contributing factor in sex differentiation is the cells from the yolk sac, which help develop both internal and external genitals.

The genitals begin to form at the embryonic stage, in weeks 8 and 9, although at that point the sex glands and organs are identical for boys and girls. A little bud develops between the legs; in boys it develops into the penis, in girls into the clitoris. Gradually

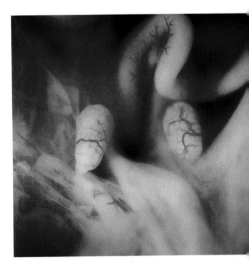

### Male genitals

In week 17 a penis has developed (left). The early testicles, which already contain immature sperm, are still in the abdominal cavity (right) and do not descend into the scrotum until late in pregnancy or sometimes even after birth.

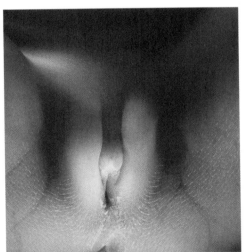

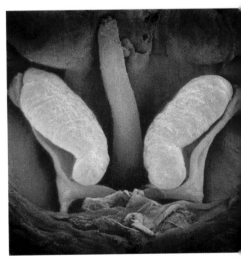

### Female genitals

By midpregnancy the clitoris of a girl baby can be discerned very clearly, owing to the as-yet-undeveloped labia (left), although in isolated cases it may be difficult with ultrasound scanning to distinguish between an incipient clitoris and a penis. The ovaries of a girl baby (right) already contain a few million immature eggs.

two little bulwarks take shape on either side of a crack. In a boy, they merge to become the scrotum, while in girls they structure the vaginal walls, and the crack will not fuse shut.

Some expectant parents are eager to know whether their baby will be a girl or a boy before the birth, while others prefer to remain in suspense. Today it is possible to determine the sex of a baby during pregnancy; most easily from about week 17 or 18 by ultrasound scan, but also by placental sampling or amniocentesis.

## Getting to be a heavy load

Most pregnant women have felt very well for several months, and pregnancy has seemed easy and fun. Now, as the end approaches, new little health problems may develop, and sometimes larger ones as well. Acid indigestion and heartburn are common, especially toward the end of pregnancy; liquid medications and pills can provide effective relief without entering the mother's bloodstream, so there is no risk of hurting the baby.

Leg cramps due to circulatory difficulties are a common problem in the second half of pregnancy. A pain that radiates from the back down into the legs may cause severe discomfort, especially in mothers-to-be who have had previous back trouble; it may be triggered by her need to balance her posture as her abdomen expands.

Varicose veins and hemorrhoids are also frequent and sometimes very aggravating complaints. Both are caused by the pressure of the enlarged uterus on the major veins that run through the pelvis. Blood has difficulty passing back through and therefore collects in the veins below the pelvis, which become swollen. This can result, in the long run, in damage to the valve system of the veins, the function of which is to prevent the blood from flowing backward.

Severe varicose veins do not disappear on their own, but as a rule they are not operated on until at least three or four months after the delivery, in hope that they will diminish on their own.

**More massage**
Many women have aching backs late in pregnancy. This is a perfect opportunity for men, who often feel like superfluous observers of the ongoing drama, to contribute in a very welcome way.

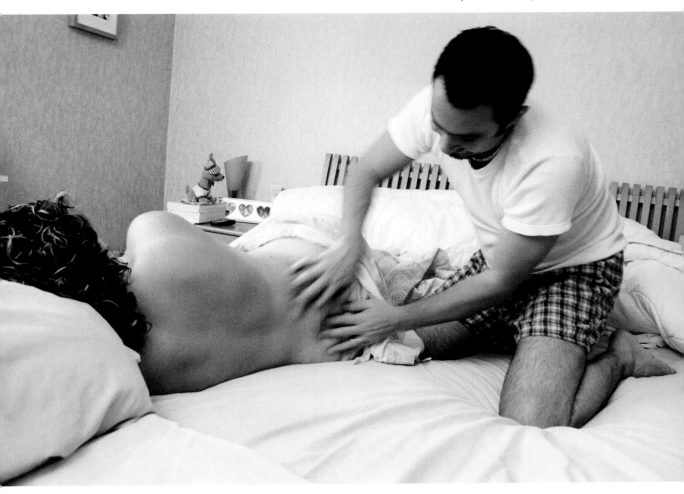

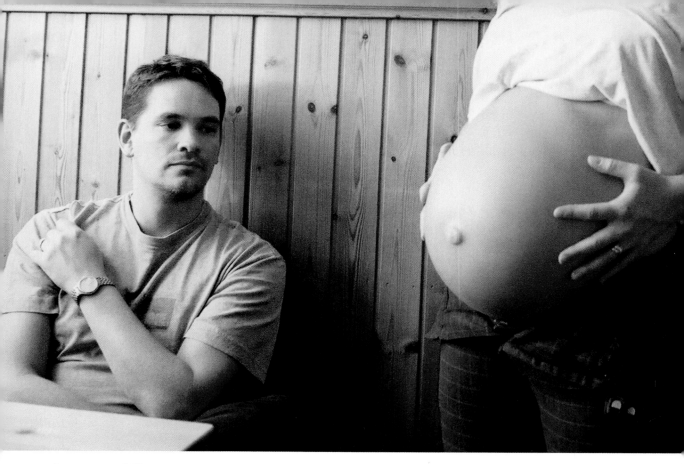

## A new way of life

Committed, supportive fathers-to-be are important during pregnancy. But no matter how overjoyed a man is about becoming a father, he may find it difficult to maintain the same level of engagement in the process as the woman. His body does not change and he will likely feel a bit left out.

For both the man and the woman, the changes brought by pregnancy can be a strain on the relationship. Often the couple's sex life is affected. A woman's sexual urge may fluctuate during the course of the pregnancy, and if it disappears altogether, the man may feel even more marginalized. For other couples, pregnancy is a time when their intimacy deepens.

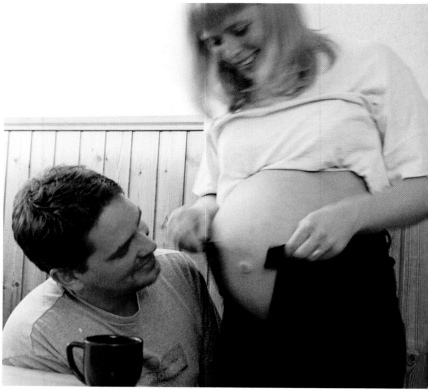

## Stress and strain

Toward the end of the pregnancy, a condition called diastasis symphysis pubis may pose a problem that is painful and impedes mobility. This is a separation of the right and left pubic bones due to a softening of the ligaments that hold them together. It makes the pelvis unstable and makes it difficult to walk. The joint may become extremely sensitive to pressure. The separation helps delivery, as it helps increase the elasticity of the birth canal. The problem usually declines within the first few weeks after delivery and generally vanishes without treatment.

Most women retain some fluid and experience some swelling during pregnancy. Moderate bloating is not a cause for concern. But serious swelling—edema—is a significant warning signal for doctors and midwives, especially if the mother is also experiencing high blood pressure and protein in the urine. Such a condition may make it is necessary to induce labor earlier than nature might have planned. The edema can affect the entire body, or it may be most pronounced in the arms and legs, making it impossible to get shoes on or rings off. These symptoms usually vanish quite soon after delivery.

In some countries pregnant women are regarded, and treated by those around them, as weak, delicate creatures. But for both physical and psychological reasons it is completely inappropriate for a woman to do nothing but rest for an entire pregnancy. In the United States and Britain, most women keep up their work until a month or less before their due date if they have no unexpected complications. Sometimes a woman may need to change some aspects of her work. For instance, a pregnant flight attendant is usually given work she can do without having to fly, and a pregnant woman who works with chemical solvents is usually reassigned. A mother-to-be should also bear in mind that it is not a good idea to do work that is very stressful. Intense stress may have negative effects on both mother and baby and is one common reason for women being given sick leave toward the end of pregnancy.

Each pregnant woman must evaluate her own work situation and check with her physician in regard to any unusual dangers in her work environment.

**Heavier by the day**

The baby is now making ever greater demands on the body of the mother-to-be. In the seventh month many women who have sailed through pregnancy so far may need to slow down a little. Most women work until late in pregnancy, but a woman in a highly stressful job, especially one with a great deal of lifting, may need to leave work earlier.

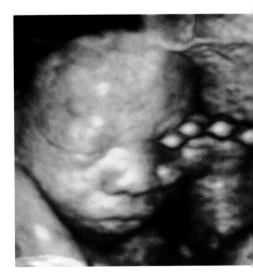

# The final spurt

One month remains of the countdown, and during this month babies usually gain a full kilogram (2.2 pounds) of their final birth weight. If the baby should be born at this point, survival is not a problem. Statistically speaking, most twins are nearing the moment of birth. Most babies (about 95 percent) are now upside down, with their heads quite a way into the pelvic canal of the mother. She becomes aware of this when she notices that her belly is now lower again and finds it easier to breathe.

## More frequent examinations

The mother-to-be now has a weekly appointment when she is weighed, her blood pressure is checked, her abdomen is measured, and her urine is tested for sugar and protein, among other things. If this is her first pregnancy, the baby's head has probably lodged quite firmly down in the pelvis. It can no longer turn over into the breech position (a problematic position that often requires a cesarean). Some babies seem to decide early on not to lie head down; in week 36 an attempt will sometimes be made at the hospital to turn such babies around, applying external pressure and massaging the mother's abdomen.

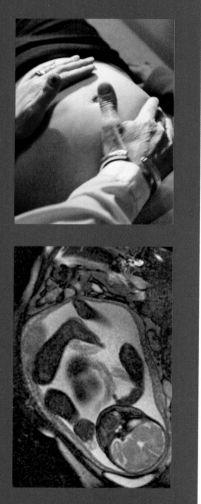

| Weeks | | 5 | | 10 | | 15 |
|---|---|---|---|---|---|---|
| Months | 1 | | 2 | | 3 | |

Approximate size
Length approx. 34 cm./13.3 in.
Weight 2000–2750 g./4.6–6 lbs.

20      25      30      35      40

5      6      7      8      9

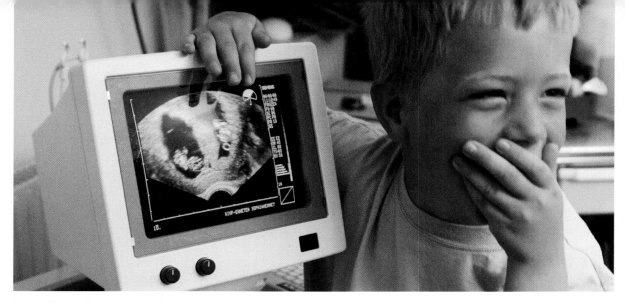

**Look, Mom—there are two!**

Suddenly two silhouettes appear on the screen. In week 10 an ultrasound scan may reveal that a family is going to increase by two new members rather than one. To a big brother, this may be exciting and thrilling. Parents, however, may need some time to adjust and feel pleased. Two babies at once—can we really cope?

## *More than one?*

Over the last two decades the new methods for treating infertility have made twin births and sometimes even triplets and quadruplets more common, since these are not always perfectly easy to control.

Without fertility drugs two to three percent of all pregnancies result in twin births. Twins may be either identical (monozygotic, uniovular) or nonidentical (biovular). When an egg is fertilized by a sperm, all the cells have the same genetic code. If, after just over a week, the egg divides into two halves, two viable embryos will develop into two individuals with precisely the same genetic code, either boys or girls. How similar they become will depend on the environment in the uterus. In some families genetic factors govern a predisposition for monozygotic twins.

For nonidentical twins the ovary releases two eggs rather than one, a somewhat more common event as a woman ages, or after fertility treatment. Of course, biovular twins may be of the same sex, but they are no more similar than any other siblings. Today genetic tests can determine with certainty whether a set of twins is monozygotic, should it be important to make this determination.

Today nearly 25 percent of in vitro fertilizations result in twins, and so obstetricians and hospital maternity units see an increasing number of them. A woman carrying twins must anticipate that her abdomen will grow more quickly and that she will have an increased risk of developing high blood pressure, edema, and anemia at an early stage. Twin pregnancies are referred to as high risk

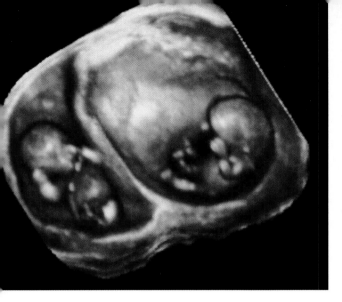

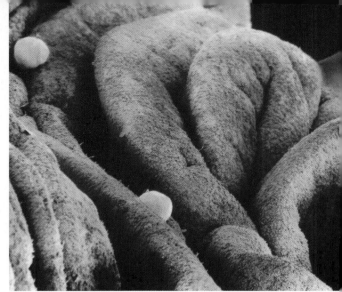

both for mother and child. The mother usually has to be examined frequently, with repeated ultrasound scans and blood flow analyses, so that everyone involved will have a good idea of how both babies are faring throughout the pregnancy.

Most women who are carrying twins find the latter part of pregnancy quite difficult. A very cumbersome abdomen makes backache a common problem, and finding a comfortable position to sleep is not easy. These women are usually anxious about the delivery. Even if the mother gets plenty of rest and does everything she is advised to do, twins tend to be born some three weeks early. There is simply no more room in the uterus, and the placentas (usually there are two) cannot go on supplying enough nutrition and oxygen for two. Newborn twins also tend to weigh less than 2.5 kilograms (5.5 pounds) on average, rather than the 3.5 kilograms (7.7 pounds) that is the mean for single births.

Risk increases when twins are born vaginally, but a mother expecting twins who has already had a normal delivery will generally be able to give birth to the twins vaginally as well. All twin births are planned and monitored with extra care. In the United States cesareans are increasingly done "just to be on the safe side," but if no unexpected complications arise, a vaginal delivery can be preferable for both mother and babies.

### Biovular twins

If two eggs are released from the ovary at once, and they both enter the Fallopian tube and are both fertilized, two embryos can develop. Biovular twins will not have identical genetic traits and may be of different sexes. They will not necessarily be more alike than any other siblings. The ultrasound image at the top left shows two amniotic sacs of the same size, containing fetuses of the same size, possibly one a boy and the other a girl.

### Identical twins

Some fertilized eggs split in two just prior to implantation in the uterine lining, and identical embryos develop (below). It is not yet known with certainty why this happens.

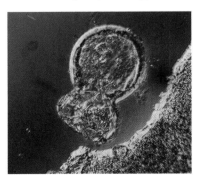

## Premature contractions

One forceful kick may spark it off: the pregnant woman's whole uterus contracts, and her abdomen feels hard as stone. These contractions last less than a minute, but they may be repeated for half an hour or more.

## Delivery time approaches

When there are only a few weeks left, time seems to drag. The baby has lodged in the pelvis, making the woman's abdomen somewhat less cumbersome and breathing slightly easier. Tingly legs are now a matter of course; the baby may kick or punch hard. The enlarging uterus puts pressure on and compresses the mother's bladder, causing her to rush to the bathroom frequently. Increased sensitivity to heat and some swelling—particularly of the feet and fingers as the day progresses—are very common. It is difficult to sleep soundly with a belly that seems in the way no matter in what position.

In the last few weeks if high blood pressure, generalized edema, and protein in the urine all arise together, sometimes accompanied by headaches, then this combination of symptoms needs to be treated, because it puts both mother and child at risk. If these symptoms become very pronounced, hospitalization may be required. If rest does not help and if, in spite of medication, blood pressure does not drop, it may be necessary, for the good of both mother and child, to induce labor. After delivery the mother's blood pressure usually falls back to normal very quickly, and her swelling also tends to disappear entirely within a week or two. Fortunately, this complication, known as preeclampsia, affects only a minority of pregnancies.

Weekly appointments with the doctor or midwife give both parties the necessary sense of security at this time. An expectant mother will have lots of questions, especially if this is her first birth. How long do I have to wait? Can't labor be induced on my due date if nothing has happened by then? Do the risks increase for a baby not born on time? At these examinations the health of both mother and child are checked, and further issues about how the delivery will be handled are explored. The expectant parents may also participate in prenatal courses to prepare for delivery. These courses emphasize how the woman can affect the course of her delivery, for instance by learning to relax and sink back into herself between contractions, and offer information about various kinds of pain relief. It is also a good idea to prepare for the delivery not running as planned. Any mother may have to be delivered by emergency cesarean section, for example, even when everything looked fine for a vaginal delivery.

Some expectant parents are able to visit the hospital maternity unit where their baby will be born a month or two prior to the due date. Knowing as much as possible before delivery increases their sense of security and eases any fears they might have about what is to come, especially with a first baby. The presence of the father-to-be in the delivery room is generally very supportive, and sometimes he is able to be of active assistance. This can endow both parents with a strong sense of togetherness.

Many men look back on the deliveries of their children as among the most fantastic experiences of their lives. But a delivery can also be stressful for a father. Some couples therefore engage a professional labor companion, known as a doula, to support them throughout the delivery. In many countries the father is present in the operating room when a cesarean section is performed. The staff must then be aware that they are responsible for yet another "patient" who might need sudden support or assistance.

**Nice and cool**

It's hard work being pregnant on a hot summer day. A pregnant woman perspires more than usual, and if the sun overheats her exposed abdomen, even her unborn child may feel the effects.

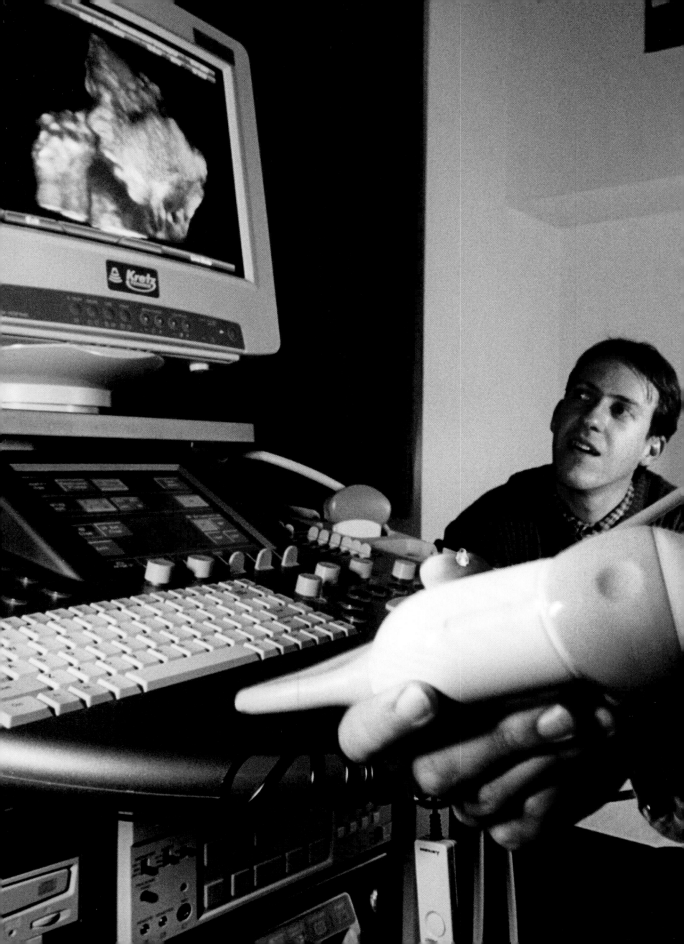

**Sneak preview**

Sometimes there will be an ultrasound scan toward the end of pregnancy, to measure the baby, particularly the circumference and diameter of the head. It is also important to know what positon the baby is lying in.

It is still virtually inconceivable to the father-to-be that there is a child in there and that he may, at any moment, become a father.

*Seeing one's unborn baby can feel like peering behind the curtain at a play that has not yet begun.*

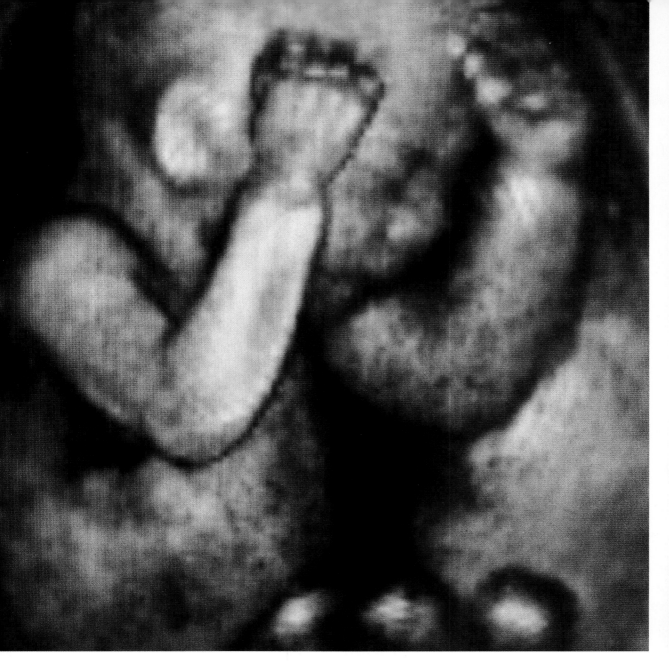

### Captured by a magical camera

The baby rests in the uterus, unaware that the outside world can peek in using new three-dimensional ultrasound. Even at this stage the expectant parents can see the unique facial features of their very own child. The photos on this page were taken in weeks 32 to 38.

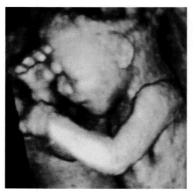

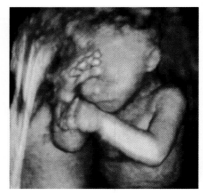

## A portrait of the unborn child

What will the baby's face look like? In the fifth week of pregnancy the embryo's face is beginning to develop, although as yet it has no eyes, nose, or mouth. Two weeks later little hollows, the beginning of eyes, have developed by the temples, and what will be nostrils have begun to show above the large opening that will be the mouth. Just a few weeks after that the face has more human features. The nose has taken shape and is separated from the mouth by the top lip. In week 16 the face really begins to look like a child's, although it is still too early to discern individual features.

The face begins to form when five projections, called processes, gradually develop under the thin layer of skin and then meet. The first process protrudes between the eyes, ending in a bay on either side, the primitive nostrils. Gradually this process becomes the nose and the center of the upper lip. Any little disruption of this development, such as a viral infection lasting a few days, may result in a cleft lip, a defect that often upsets parents terribly but that can be surgically corrected with perfect results. Two other processes shoot out from the sides of the head, one under each eye, to form the cheeks and the sides of the upper lip, and the last two develop below the mouth, forming the lower lip and chin. Muscles then develop and grow from this framework and are supplied with blood vessels and nerves. Soon the face can move, and the first facial expressions occur. But there is still a long way to go from this early formation to the unique and highly expressive human face that takes shape by the end of pregnancy.

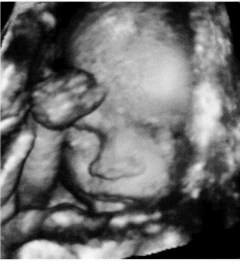

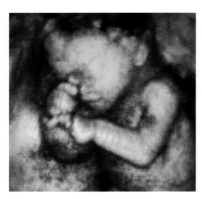

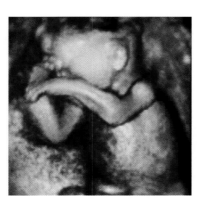

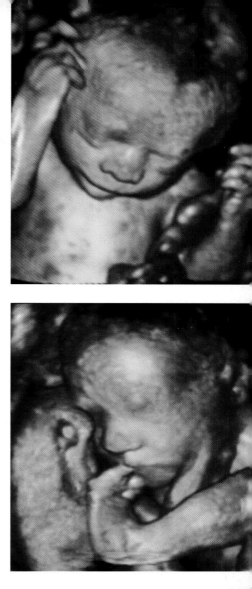

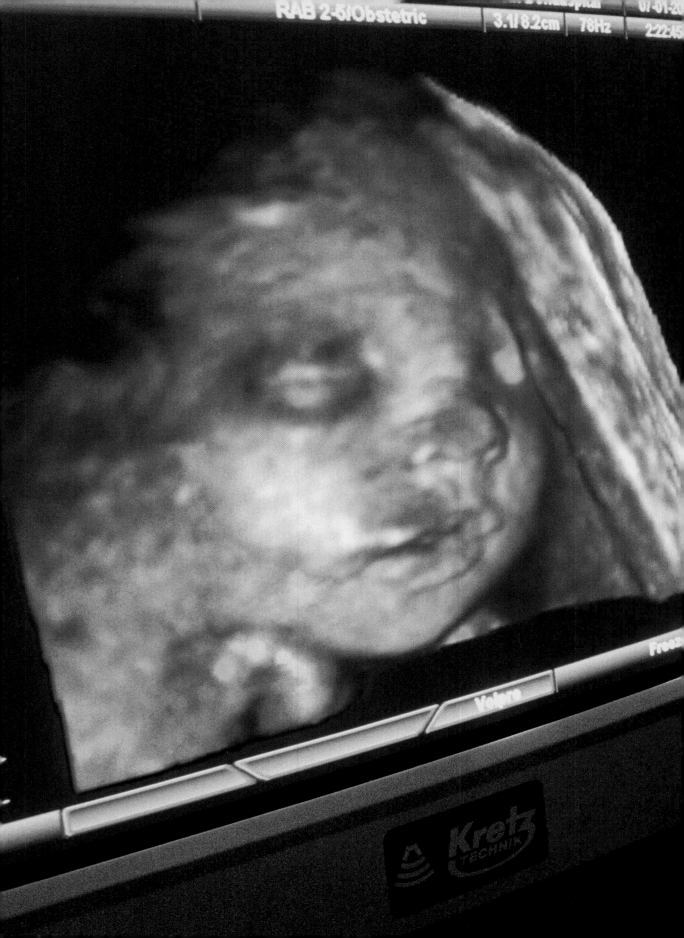

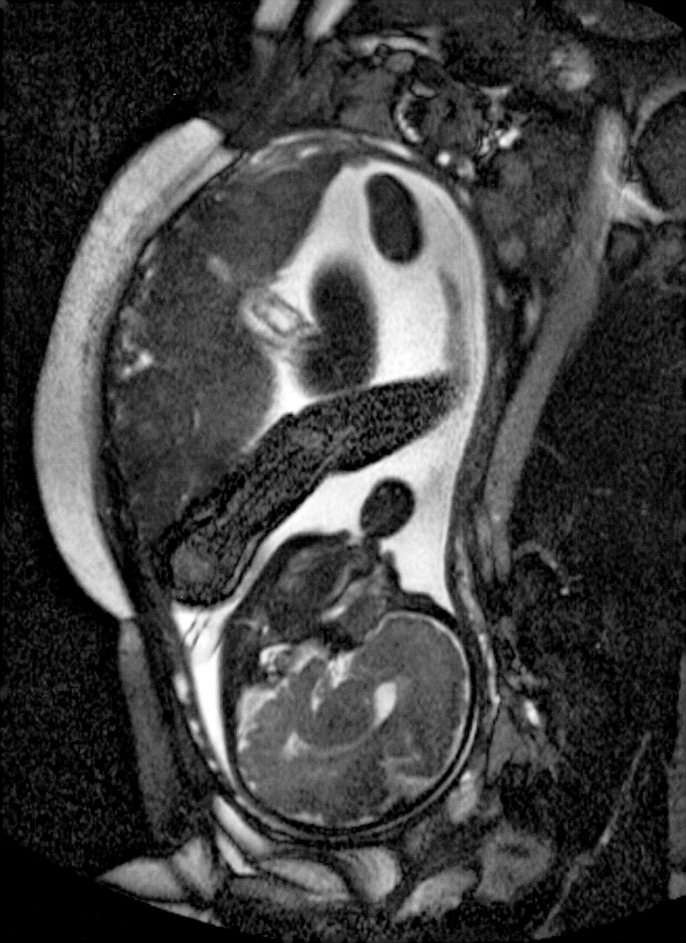

## Overdue

If labor does not begin on its own, the outer time limit that is not allowed to pass is forty-two weeks. After this time the placenta may not be able to provide the growing child with sufficient nutrition, and the amount of amniotic fluid usually decreases, restricting the fetus's freedom of movement.

A pregnant woman who has passed her due date has frequent appointments with the doctor or midwife, and if the baby is not born on its own after forty-two full weeks, there is nothing to be gained by additional waiting. The woman then goes to the hospital, and her labor is induced. Old remedies such as castor oil and enemas have been replaced by more reliable ways of inducing contractions, including medications given by nasal spray and intravenous drip. Hormone gel (prostaglandins) is usually applied to the cervix to cause it to ripen and let labor proceed unimpeded. If it has begun to dilate on its own, the fetal membranes may be ruptured, so the amniotic fluid begins to leak out. This often speeds up the contractions, because as the baby descends farther into the birth canal, its weight stretches the muscles of the cervix. If all efforts to induce labor should fail, the baby is delivered by cesarean section.

Babies who have been carried past term often have a characteristic appearance for the first few days, with wrinkly, quite dry skin. But this odd elderly look soon disappears, and their newborn appearance and sweet baby look shine through.

Many expectant mothers find going past the estimated due date more trying than almost any other aspect of pregnancy. Their stomach is heavy and uncomfortable, they feel impatient, and may have to field daily queries from anxious grandparents and others. Still, most gynecologists and obstetricians all over the world agree that when it comes to delivering a baby, the wisest thing is to be prudent, because a baby who spends some extra time in the uterus often has a good reason for doing so.

**What's taking you so long?**
Sometimes the baby seems to be reluctant to face the world, and the due date passes uneventfully. The image opposite is taken by use of Nuclear Magnetic Resonance (NMR), a technique which unlike X-rays is completely harmless for mother and child.

# Labor and Delivery

**The baby, exhausted after the strain of coming into the world, cuddles at mother's breast. Billions of women have given birth, and still every birth is unique–simultaneously a routine event and a miracle.**

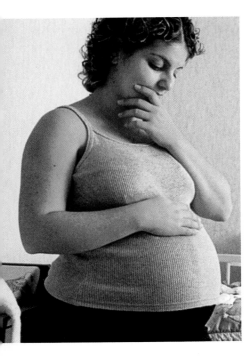

## The time has come

Almost nine months have passed. The due date is approaching, and for the parents-to-be, every day is full of anticipation. The baby might come at any time, but how will they know when? There are three common signs that birth may be near at hand: regular contractions of the uterus, the breaking of the waters, and bloody discharges.

Most women have occasional contractions late in pregnancy. For a moment the uterus becomes hard as a ball, only to relax again. Contractions that become more frequent, returning time after time and with increasing intensity, may be a sign that the labor has finally begun. When the contractions come at four-minute intervals, it is wise to head for the maternity unit, although if the hospital is far away, it may be advisable to leave earlier. Sometimes the contractions stop on the way, and after an obstetrician or a midwife examines the woman at the hospital, she is sent home again. This is perfectly natural, if the medical staff estimate that it may take several days or even a week for the real contractions to begin. Trying to induce labor with labor-inducing drugs (oxytocin) administered as an intravenous drip may have complications. It is generally done only when there is a specific reason that the child needs to come into the world as soon as possible.

Sometimes the first sign that the delivery has begun is that amniotic fluid runs out of the vagina. If this happens, it is time to go to the hospital, even if contractions have not begun. Sometimes, however, a little leak of urine in response to the baby's kicking in the direction of the mother's bladder may be mistaken for the breaking of the waters. The mucus plug may also dissolve and be expressed as a heavy discharge. If the amniotic fluid is cloudy from the contents of the baby's intestines, this may (but not always) indicate that the baby is no longer faring very well in the uterus. When the amniotic sac ruptures, the risk of infection increases. In that case the baby should be delivered within a day or two.

Any bright red bleeding, however large or small, is an urgent reason for the woman to go straight to the hospital. Parts of the placenta may have torn lose from the uterine lining, which jeopardizes the baby's oxygen and nutrient supply. A small amount of blood mixed with mucus is the most common type of bleeding and is attributable to the maturing of the cervix, as it pulls up and becomes wider in anticipation of the birth. The mucus plug that has been blocking the cervical canal is usually tinged with blood and may come loose during the small preparatory contractions. Intercourse may also cause bleeding. Although in the past it was believed that intercourse late in pregnancy (known as "one for the road") was a good way of triggering the delivery process, today this notion is considered a myth.

### Off to the hospital

Most pregnant women have been told precisely when to go to the hospital, but uncertainty often causes many to arrive earlier than is really necessary. Take it easy! There is seldom any terrible rush.

## What makes labor begin?

Most babies are born sometime between week 38 and week 42. The beginning of contractions is governed by a number of different factors interacting in the body, particularly hormones. A hard kick from the baby is also thought to trigger the process. The progesterone that builds up in the placenta during the course of the pregnancy, increasing month by month, has various roles, one of which is to ensure that the muscles in the wall of the uterus remain relaxed and calm. Three other groups of hormones produce the opposite effect: oxytocin, prostaglandins, and cortisone. For the birth to begin, the amount of progesterone in the bloodstream has to decrease, although it is still not known precisely what causes this to happen sometime around week 40.

If labor must be induced, oxytocin is the most commonly used hormone today. It is administered as an intravenous drip. Prostaglandins soften up a cervix that is unripe, preparing it to cope with the contractions. When an induced labor is planned in advance, prostaglandins are administered as vaginal suppositories or gel for one or more days prior to delivery.

**Up and about**

Moving around between contractions, ideally with some kind of support, feels good and makes for more efficient labor during the early phase. Gradually, however, the contractions become so strong that many women prefer to lie down.

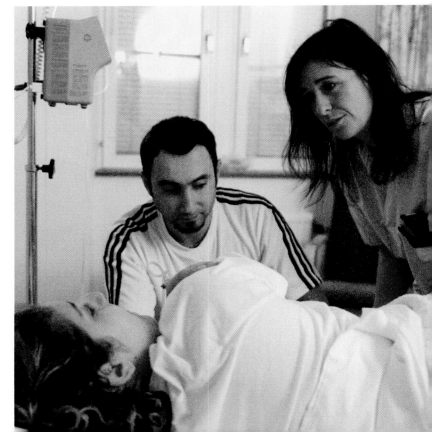

## The opening of the cervix

Upon arrival at the hospital, and after a short admissions procedure and a shower, the mother is usually examined, to give both the staff and the parents some idea of how far the first stage of labor has progressed.

During this first stage regular contractions dilate the cervix, which becomes shorter and is then effaced altogether, so as to allow the baby to pass. The cervix opens to 10 centimeters (3.9 inches) when it is fully dilated. This phase is the longest stage of labor and may take up to fifteen or twenty hours, although in women who have given birth previously it often progresses much faster. During this phase the head (or bottom) of the baby also presses, with rotating movements, down on the pelvic floor.

During the first few hours it is good for the woman to get up and walk around and then to rest. Many women feel pain briefly in the middle of each contraction. Standing up and walking around may help the contractions do their work, since the law of gravity helps the head of the baby press down into the cervix and the pelvic floor.

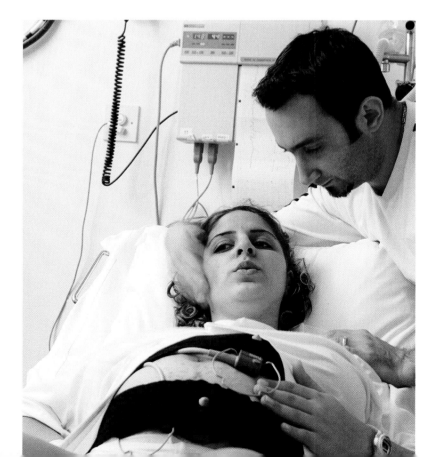

**Close contractions**

When the contractions are only a few minutes apart, many women ask for some form of pain relief. At this time the father and a labor companion can make important contributions. Their very presence and support have a calming effect. A belt to monitor the baby's heartbeat also provides reassurance.

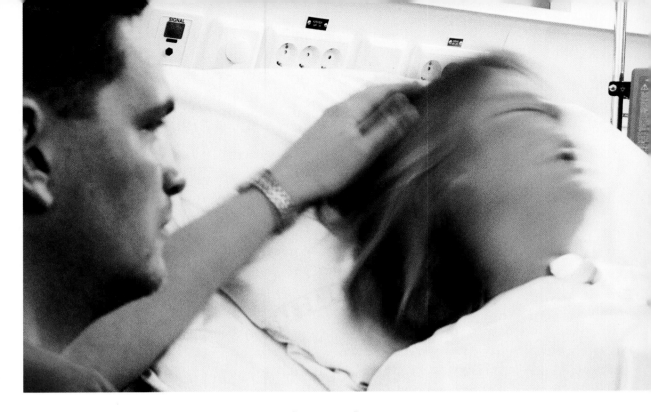

## Does it have to hurt?

Historically, giving birth has been intimately associated with pain, at least for the human species. Contractions are, in fact, traditionally referred to as "labor pains"; the biblical "In pain you shall bring forth children" (Genesis 3:16) links birth and pain. This particular kind of pain, however, is different from almost any other. Moreover, most women forget their pain at the very moment their child is born. The result of the "labor pains," after all, is a perfectly unique baby of one's very own.

Maternity care has long made available various kinds of pain relievers that are effective for the mother, usually without risk to the baby or to the contraction process. But in recent years expectant mothers have desired to get away from medical and technological solutions and to work in a more natural way with the body. With support and preparation women are more inclined to tolerate a certain amount of pain. Still, pain perception is highly personal, and those caring for the mother must be receptive to her needs. People always experience pain as more intense if they are frightened. It is thus very important that the birthing mother always be very well informed of what is happening at each stage.

# Pain relief

## Epidural anesthesia

Epidural anesthesia is a combination of rapid pain-relieving medication and an anesthetic that blocks all the nerves issuing from the lower spinal cord. One type that is gaining favor is called a walking epidural, because the woman is still able to use her legs. An epidural block may be administered only by an anesthesiologist. Too much anesthetic can affect the mother's blood pressure as well as her breathing. This could, in turn, have negative effects on the baby. Sometimes an epidural can cause the expulsion phase to be prolonged, making it necessary to deliver the baby by suction or forceps. The woman may also develop an itching from the medication, or have trouble urinating for a short time after the birth.

## Cervical anesthesia

Local anesthesia of the nerves around the cervix (a paracervical block) often provides effective pain relief during the period when the cervix is widening, and the anesthesia itself may accelerate the dilation process. There is, however, a certain risk that the baby will be affected, and therefore physicians are somewhat restrictive about using this method.

## Nitrous oxide

A once-very-common type of pain relief is the inhalation of a mixture of nitrous oxide and oxygen. The woman breathes through a mask she holds up to her own nose and mouth when she feels a contraction coming on. In the respites between the pains, she breathes normally, without the mask. In some women the intoxicating effect of this mixture may be anxiety-producing.

## Analgesics

If the woman is tired or worn out, she may sometimes be given an analgesic such as mependine or fentanyl intravenously or by injection. These drugs are used with caution, however, because they can affect the alertness of the baby and depress its cardiac activity and breathing.

## Pelvic anesthesia

Local anesthesia of the pelvic floor, known as a pudendal block, during the final stage of labor, was once one of the most common types of anesthesia. Today it is used mainly when the delivery has to be completed with suction or forceps, or if an episiotomy or stitching of the perineum is necessary.

## Acupuncture and hypnosis

Acupuncture (in which needles are inserted at specific pressure points) and hypnosis are other types of pain relief available at certain hospital maternity units. Recent years have seen less skepticism toward these types of treatment, and many women are prepared to bear witness to their effectiveness.

## Saline solution and TNS

Subcutaneous injection of small amounts of a saline solution, or the administration of tiny shocks through the skin (transcutaneous nerve stimulation, or TNS) appears to stimulate the endorphin system (the body's own pain-killers) without negative side effects.

## Massage and heat treatment

These methods have been in use since time immemorial. Hand massage is excellent, and various ways of applying heat can relax and soothe the woman in labor.

## Confidence in the care provided

Fear and pain are intimately interrelated. A safe, friendly, expert environment may be as effective as many kinds of pain relief. The presence and participation of the father-to-be and the continuous support of a doula (labor companion) have been shown to reduce the need for medical pain relief.

## Free choice of position

Some women choose to give birth lying on their backs, others squatting, kneeling, or sitting. There are no rules, no absolutes, and no obvious choices. When the expulsion phase approaches, the mother usually senses what will be best for her.

## *Time to bear down at last*

The second part of the birthing process is called the expulsion phase, and it lasts from the moment the baby's head or bottom presses down on the pelvic floor until it is born. The mother often feels a sudden surge of pressure on her rectum, after which the bearing-down reflex takes over. During this stage it is important for the mother to bear down actively when the reflex is activated during contractions, and to relax as much as possible between contractions, breathing deeply and calmly. Some women find it a relief to bear down, while others find this an extremely painful part of labor and are upset by the force of the reflex.

Looking back a hundred years to home births, babies were probably delivered in many different ways, with the mothers in many varied positions. Today in the developed world most babies are born in the hospital, as this has come to be considered safer. In the hospital it is easier to monitor the labor process and the cardiac activity of the fetus, as well as its precise position. Under these conditions the woman is more likely to spend her labor

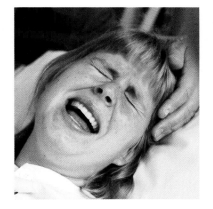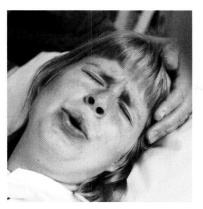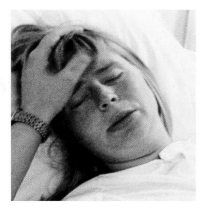

in bed, where she delivers in a semireclining position.

In the last few years, even in hospital births, the trend has reversed toward allowing women a greater degree of freedom in the choice of labor position. Opinions remain divided, however, regarding the safest, least painful, and most natural kind of childbirth from the perspective of both the expectant mother and her baby.

**Bear down, persevere, rest**

When it is time to bear down, the doctor or midwife gives very definite instructions and may even seem a little gruff. It is very important that the woman giving birth and her caregivers communicate well. This will make the second, pushing stage as short as possible and prevent injury to the woman's perineum and rectal muscle.

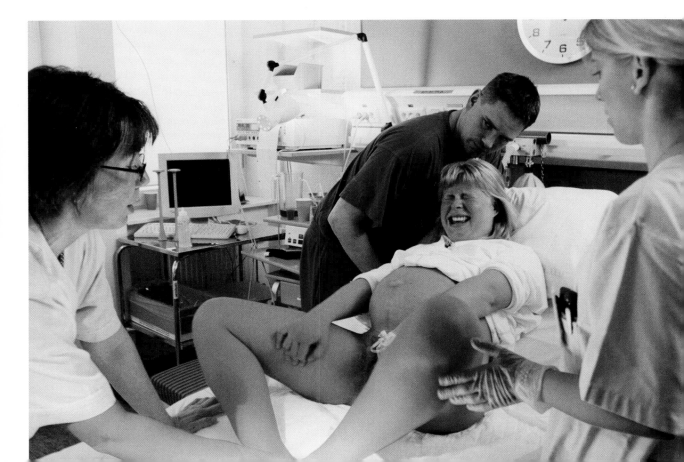

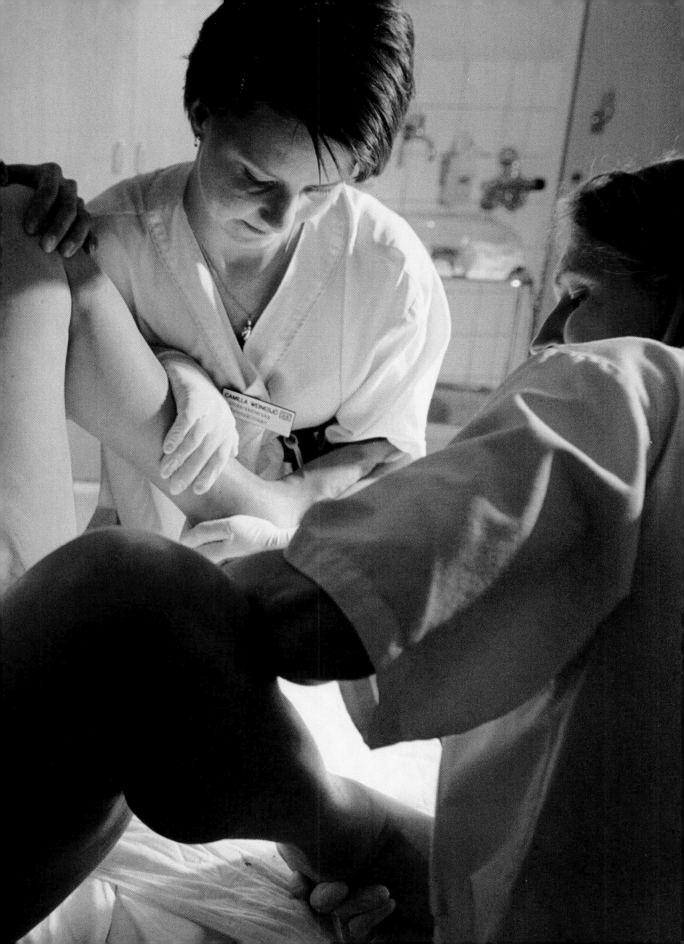

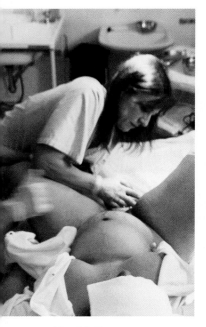 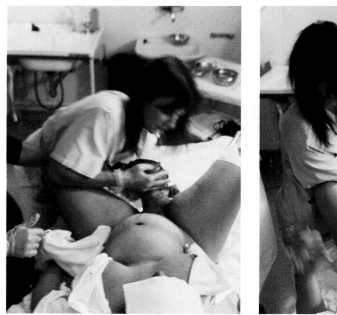 

## The birth

The head has pushed down toward the vaginal opening. When it is partly visible, the midwife (above) gently probes for the baby's chin, where she can get a firm hold. The body may need to be slightly rotated. A good grasp of the head will bring the rest of the baby out in just seconds (opposite).

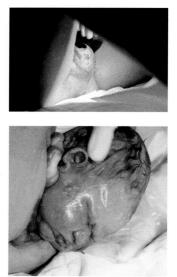

## *Baby on the way!*

Sometimes the baby will be born after just a few strong contractions, but often the expulsion stage is slower. Usually it takes less than a couple of hours, although women giving birth for the first time may need more patience. At this point cooperation among all parties—mother, father, and doctor or midwife—is essential.

A small incision in the perineum between the vaginal opening and the anus is sometimes done to accelerate the birth and/or to reduce the risk of tearing in the perineum. But episiotomies, as they are called, are done less frequently today because women who tear on their own have been shown to do better generally than those who have had an episiotomy. Another technique used to speed birth is vacuum extraction, using a suction cup made either of metal or of a rubberlike material. With the help of negative pressure from a pump, the suction cup is placed on the baby's head, and the physician or the midwife then slowly and methodically pulls, in time with the contractions and in the direction of the birth canal. Babies delivered this way often have a cup mark for a few weeks, but suction is generally considered harmless. An alternative aid is the obstetrical forceps, with two blades that can grasp the head without injuring the baby.

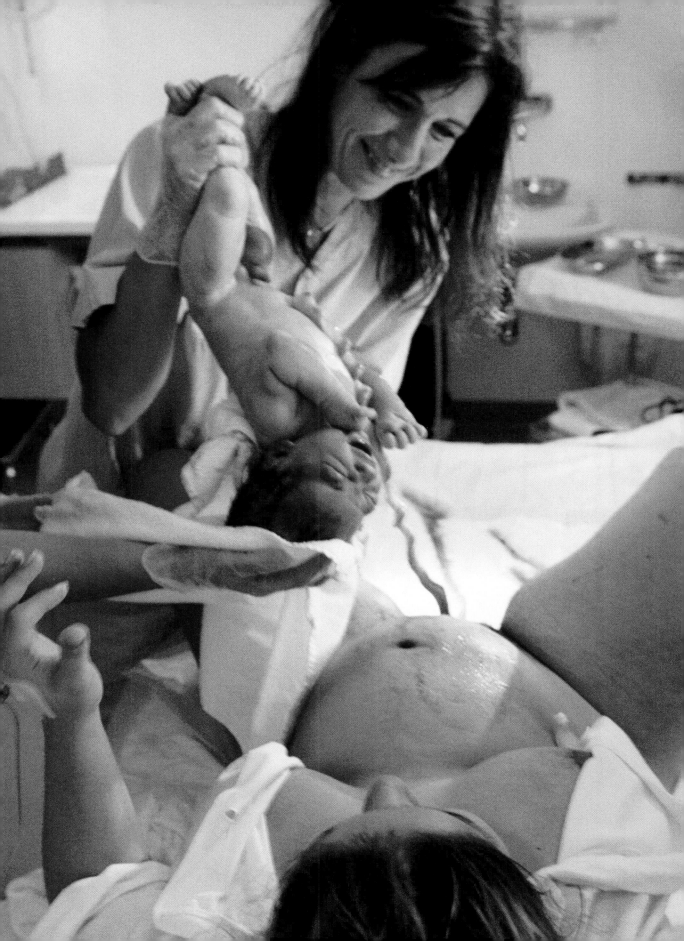

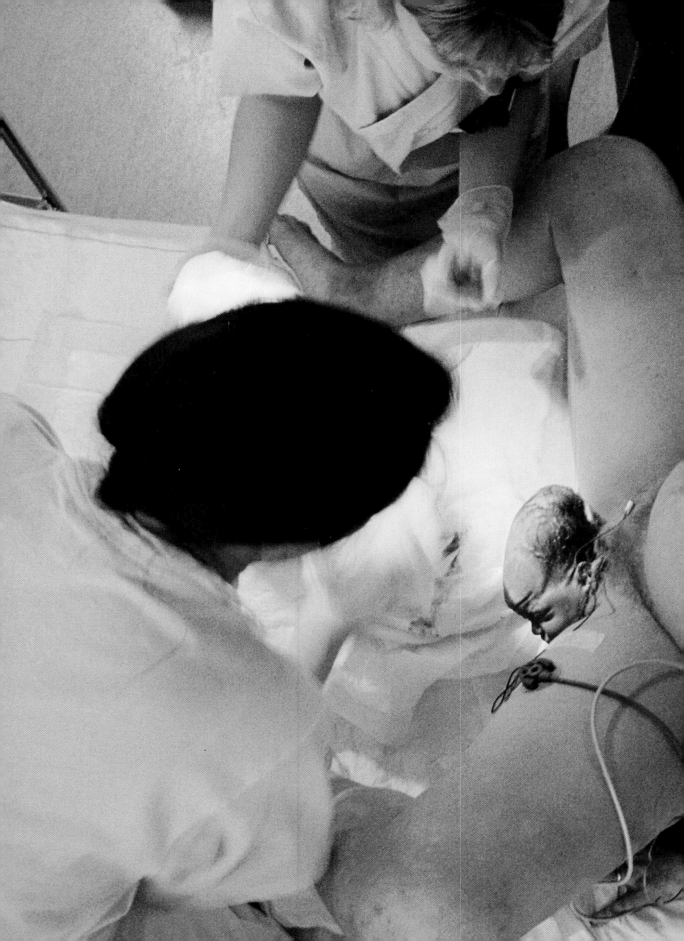

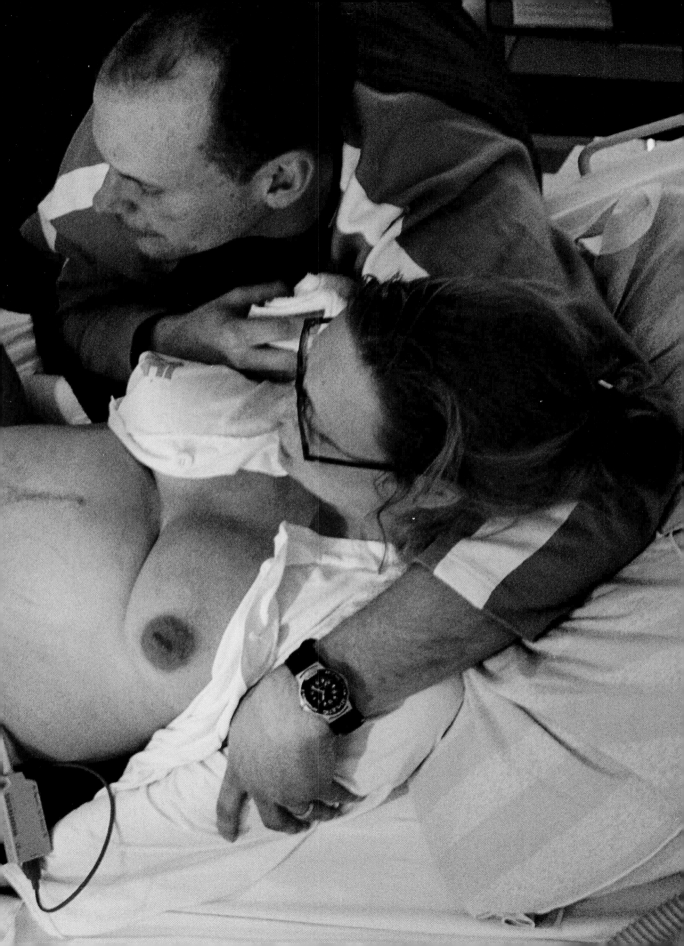

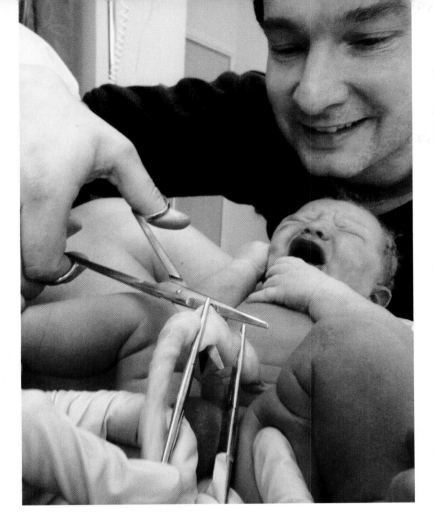

**Grand opening**

Cutting the umbilical cord is a fantastic experience for many fathers (left). For the baby, it means surviving independently for the very first time. Nine months of intimate unity with the mother are now severed, replaced by new emotional ties.

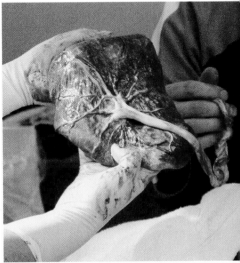

**Checking the placenta**

It is important for the midwife to check that there do not seem to be any abnormalities in the placenta and that it is delivered whole (above). If any fragment should remain in the uterus, it could cause relatively serious bleeding for several weeks after the delivery.

## Welcome to the world

The moment when the umbilical cord is cut and clamped is an indescribably critical time, for the baby's lungs undergo their very first test. The baby's first cry draws air into the pulmonary cavities, and the crying and coughing reflex release whatever mucus may be in there. Sometimes the phlegm has to be brought up through a tube. The doctor makes careful notes regarding the baby's first breaths, skin tone, and muscle strength.

Although the mother tends to be preoccupied by what is going on with her baby, her body still must complete the afterbirth stage: the discharge of the placenta and the fetal sac. This may take from a few minutes to an hour. If the baby is put to the mother's breast, the placenta exits more readily. The mother is then checked for any serious tearing of the perineum or between her labia. Little tears will heal on their own, while more serious ones require stitches.

### Exhaustion

After many hours of laborious effort and stress during the delivery, the baby encounters the world. Everything is new—air, light, noise. The mother's body provides warmth and security in the midst of all the confusion.

## *What the delivery is like for the baby*

Even when there is no need for suction or forceps, being born is extremely stressful for the baby who, with every contraction, is squeezed so tightly together with the placenta and the umbilical cord that some of the oxygen supply is cut off.

Monitoring the baby with a scalp electrode (known as cardiotocography or CTG) shows how fast the heart is beating at any given moment. The heart rate is usually reduced, particularly during contractions late in the delivery, but if the baby is doing well, it quickly picks up again, and the oxygen supply is also normalized. If the pattern of the CTG curve changes, or if the baby's heart rate remains slow between contractions, extra measures may be necessary. The mother may be placed on her side or given oxygen, or if she has been receiving medications to speed contractions (oxytocin), the amount may have to be decreased.

Babies have a fantastic ability to cope with the stresses of birth. The adrenal glands excrete large amounts of both adrenaline and noradrenaline into the circulatory system, keeping the heart rate up and making it easier for the heart to pump. This improves the blood, oxygen, and nutrient supply to the brain. Never again in a human being's life do so many stress hormones flow in the blood at once—it is a feat to be born!

Hormones also play a very important role in preparing the lungs for life outside the womb. Adrenaline reduces the accumulation of fluid that filled the cavities of the lungs in utero. One sign that the lungs are working properly is that the skin of the newborn becomes rosy, the muscles tense, and the cries get louder and harder. This is the moment when all the things the baby has been preparing for in the womb are put to the test.

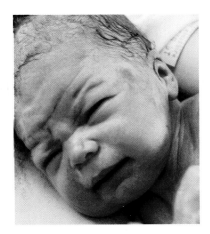

**A few minutes old**
The baby squints, and adrenaline pumps through the little body, triggering the breathing process. It will be some time before the baby can take in all the new impressions.

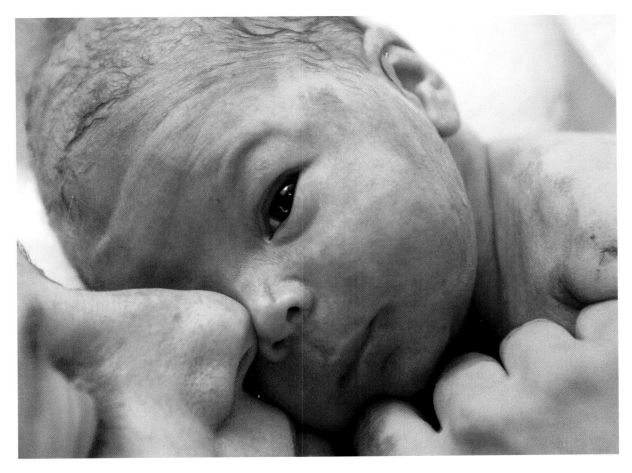

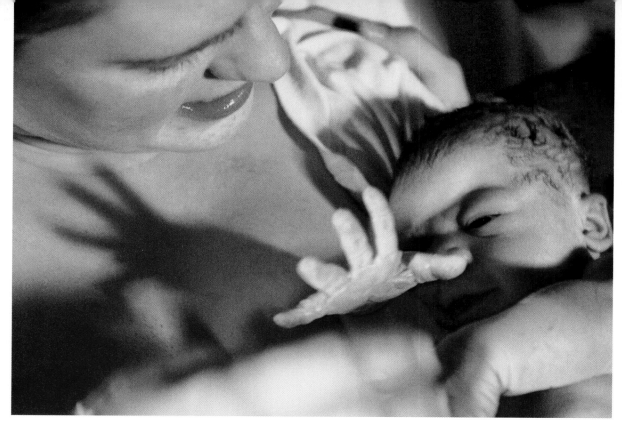

## Magical moments

At first the baby is wide awake, owing not least to the enormous amounts of supplementary adrenaline its adrenal glands have been producing, but it soon falls asleep at the breast, worn out. Activity gives way to calm, chaos to harmony. The world vanishes far away.

## *Aren't you lovely!*

The delivery is over. On the mother's breast lies a little newborn, warm and alive against her skin. All the parents' fears are past, and the mother's pains are over. Now it is time to take a deep breath and for baby and parents to begin, slowly and respectfully, to become acquainted. Today hospitals are more sensitive than in the past to the importance of giving the new parents the peace and quiet they need with their baby immediately after delivery. Today the staff is prepared to wait for a while before bathing, weighing, and measuring the baby.

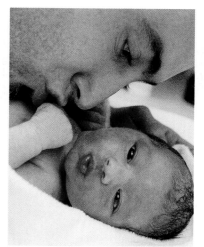

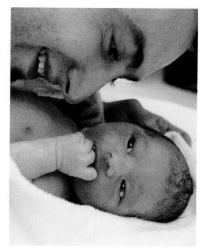

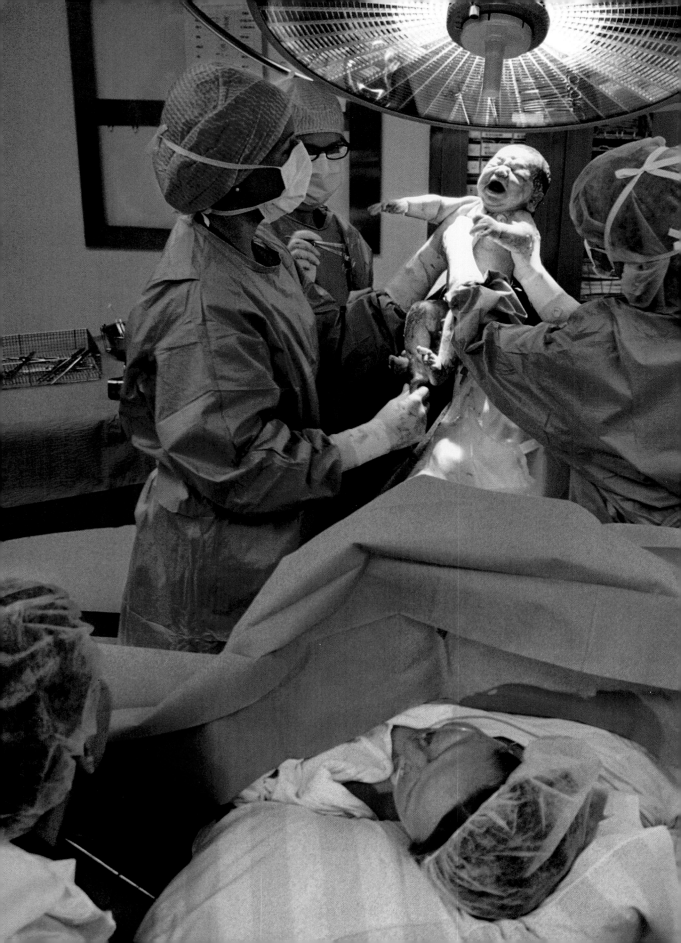

# C-sectioning into life

In recent years the proportion of cesarean births has increased dramatically. In the United States it is now between 20 and 25 percent. There has even been a debate regarding whether it should be a woman's right to choose a cesarean. Unless it is an emergency, this decision is generally made by physician and patient in consultation.

A cesarean may be planned well in advance or may become necessary during labor because the baby shows signs of doing poorly in the womb, such as from a sudden reduction of the blood supply and consequently of the oxygenation of all tissue. A cesarean may also be done when the contractions are very weak, when part of the placenta separates from the uterine wall too early, or when the umbilical cord has slipped into the vagina where the baby's head can compress it. Cesareans may be planned when the baby is too large in relation to the birth canal or is in the breech position, feet down, or, in very occasional cases, in transverse position. Serious illness in the mother may also be a reason. Fear of childbirth has contributed to the increasing number of cesareans performed today.

In emergency cesareans general anesthesia is sometimes used, but in most other cases the mother is given spinal anesthesia, which eliminates all sensations of pain in the lower half of the body. The mother then remains fully conscious at the moment of birth.

When a surgeon performs a cesarean, he usually makes a horizontal incision relatively low on the abdomen. He then opens the uterine wall, and when the fetal membranes are ruptured, the amniotic fluid rushes out. The child is carefully lifted out, and the umbilical cord is cut. Then the placenta is removed, and the uterus and abdominal wall are stitched. The entire operation usually takes no more than half an hour, and in a real emergency the part of the operation that rescues the baby can be done in a matter of minutes. A mother who is delivered by c-section spends a few extra days in the hospital. After returning home, and for a few weeks, she will experience certain effects of the surgery and will need help. She will be more tired and less free to lift than a mother who has had a vaginal delivery.

**Birth celebration**

For many women who undergo a cesarean, being awake at the moment of delivery is very important. Today most mothers can be awake during the operation and can hold their newborns just a few minutes after the birth. The father is often there to share the joyful moment.

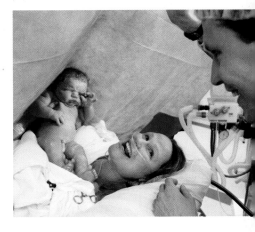

**Mother's milk**

Putting a baby to the breast immediately after the delivery is very important in triggering milk production. The first milk is called colostrum. After a few days the mature breast milk begins to flow—often quite copiously!

## The first feeding

Immediately after the delivery the baby is placed on the mother's belly and often finds his or her way to the breast for a first feeding. Most babies have been practicing sucking on their fingers and toes in the womb, but sometimes both mother and baby require a little extra instruction. The baby's sucking is actually needed for the milk production and flow to begin. Nerve reflexes from the nipples are transmitted to special centers in the mother's lower brain, from which signals go to the pituitary gland, which begins to excrete prolactin, a hormone essential to milk production. Another hormone, oxytocin, affects the mammary glands, so that the milk is pressed out into the breasts. Oxytocin also helps the uterus to contract; the mother may notice contractions while breast feeding, particularly soon after delivery.

For the first few days after the delivery, the breasts are tender and swollen. The first milk is known as colostrum. Only a very small amount of colostrum is produced, but these drops are very precious to the baby, particularly in terms of early immune defense reactions. After two to three days the actual breast milk begins to flow. The earlier the child becomes used to taking the breast, the more quickly milk production will pick up.

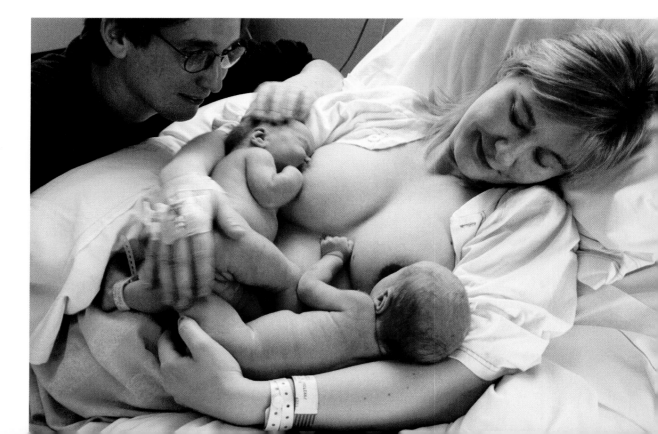

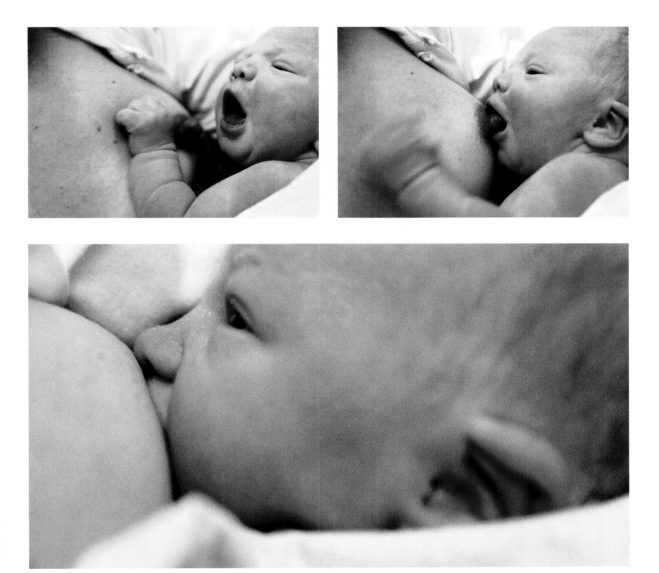

## Breast-feeding is more than just feeding

Today it is known with certainty that breast-feeding is important, at least for the first few months of life. It creates intimate contact between mother and child, and breast milk is the perfect food for babies, full of important nutrients and disease-fighting antibodies. Breast milk also contains substances that have a calming effect on the baby. When a mother nurses, oxytocin, a calming hormone, is released in her body, making her feel more harmonious as well.

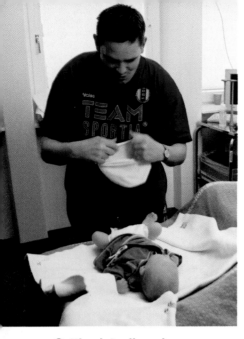

## Stepping out into life

Before discharge a pediatrician examines the baby. He or she checks the muscle tone and reflexes—for example, the baby's walking reflexes—because this reveals a great deal about the functioning of the baby's brain and nervous system. The fontanels, the vertebral column, and the hip joints are examined with particular care.

Most newborns have a slightly yellow tinge to their skin, owing to buildup of a chemical called bilirubin from blood cells that are no longer required. If the yellow tinge is more than moderate (infant jaundice), the baby receives light treatments (phototherapy) until the bilirubin is reduced. When the doctor has verified that everything is in order, the new family is released to go home.

**Getting into diapering**

It soon becomes very routine, but the first diaper change is a ceremonial occasion (above).

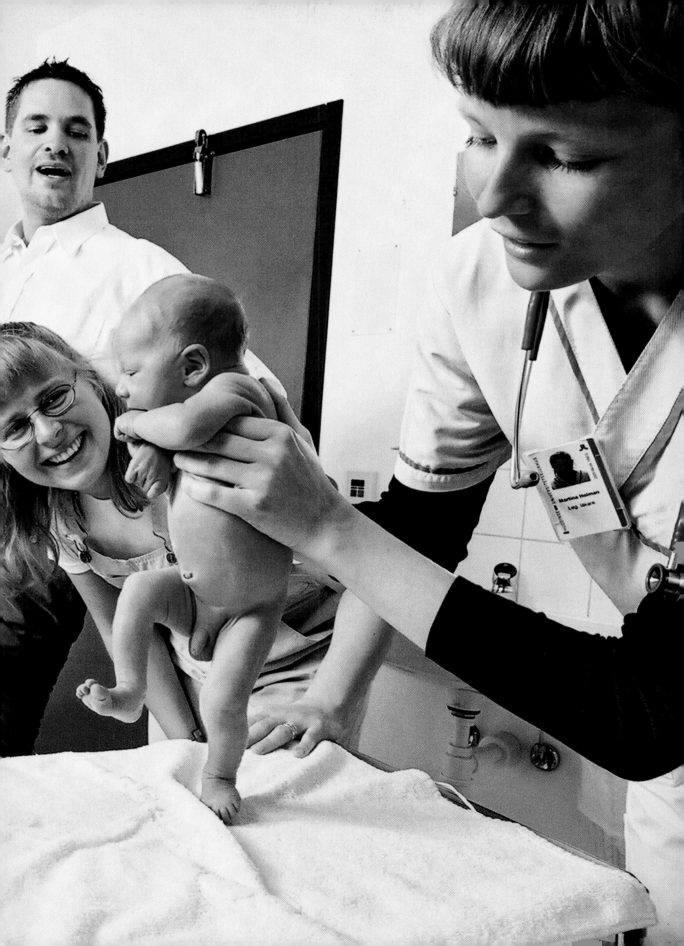

# When nature needs help

**Although most couples are capable of having one or more children, infertility is a large and growing problem. Estimates indicate that every seventh or eighth couple has more or less serious difficulty becoming pregnant.**

## Why don't we get pregnant?

There are many explanations for involuntary childlessness. Some are well known and clearly defined, while others are more diffuse and ambiguous. The latter category includes environmental toxins and radioactivity as well as the stress of contemporary life. In individual cases pinpointing the less tangible explanations is difficult, so doctors tend to work their way first through the well-known medical causes of reduced fertility or infertility. Such medical examinations may be both time-consuming and expensive and do not always provide a useful answer.

In an estimated 40 percent of cases of childlessness the problem lies with the man only; in another 40 percent with the woman only; and in about 20 percent the problem has to do with the couple. This latter category includes "unexplained" infertility, meaning cases where no problem is found in either partner but the woman still fails to become pregnant.

The main causes of female infertility are either mechanical or hormonal. The most common mechanical cause is scarring and blockage from a uterine or ovarian infection (sexually transmitted diseases) or, less commonly, as a consequence of surgery, including a childhood appendectomy. Sometimes, although relatively seldom, there may be abnormalities of the cervix, the uterus, or the Fallopian tubes. One hormonal cause may be that the pituitary gland does not excrete the proper amounts of gonadotrophin, the hormone that stimulates the growth of follicles in the ovaries and trigger ovulation. Sometimes a high level of stress hormones prevents the woman from becoming pregnant. The hormonal balance may also be disturbed if the woman is seriously overweight or underweight (for instance, if she has anorexia).

Causes of male infertility may include overweight, overexertion in sports, or alcohol or substance abuse. Malformations and impotence may be involved, but in contrast to female infertility, the reasons for male infertility often remain obscure.

## Examination and testing may provide answers

A relatively young, healthy couple in which the woman has regular menstrual periods will not usually be offered any medical examination at all until the couple has been trying to become pregnant for a year.

When the examinations begin, both the woman and the man are checked for any possible infections, particularly bacterial infections such as syphilis, chlamydia, and gonorrhea, but also viral infections such as hepatitis and HIV. Bacterial infections can easily be treated with antibiotics, while viruses may be chronic and unresponsive to treatment. Sometimes special hormone analyses can provide valuable information. The interior of the uterus can be evaluated by direct observation using an endoscope.

### The decision is made

"We've been trying for almost two years, but there's been no pregnancy. Are we doing something wrong? Is one of us unwell?" Asking for help may be difficult, but a childless couple must not lose valuable time, so turning to a specialist is the right move. The doctor can begin to investigate the situation right away, which gives the couple the peace of mind that they will need eventually to succeed. They can turn their worries over to the experts, and today there is a very large probability that the woman will be able to become pregnant.

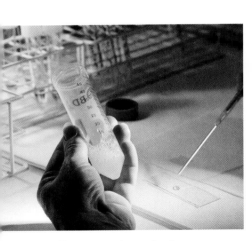

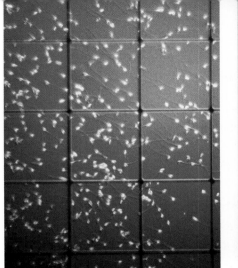

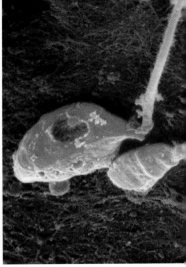

**Sperm–a closer look**

Thousands of madly swimming sperm compete for space in a counter chamber (above center). The man has submitted a sperm sample, and the laboratory staff can determine its quantity, quality, and mobility relatively quickly under a microscope. Sometimes they find a very low sperm count, or the sperm may display aberrations or impaired mobility (above right).

Sperm can also be removed directly from one of the meandering canals in the testicles if the ejaculate contains no sperm at all (below).

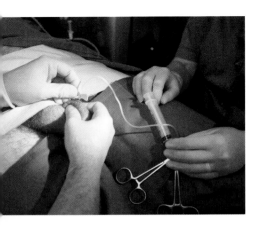

## *Are the sperm viable?*

A sperm test includes a sperm count and an investigation of the sperm's mobility and appearance. Many sperm have tiny defects, and a detailed microscopic examination may find that only about 10 percent of the sperm are absolutely perfect—even in a completely healthy, fully fertile man. Only if the number of sperm in an ejaculation is below five or ten million, and if some of them have poor mobility, is there good cause to suspect male infertility.

Sometimes adopting a healthier lifestyle may enhance sperm production, but even if a man stops smoking and drinking, loses weight, and begins to exercise, the quality and quantity of his sperm seldom improve radically. Drug abuse or anabolic steroids may sometimes explain poor sperm quality. Generally, however, the cause cannot be identified. The damage may have been done long ago, perhaps even in the womb. Today, in the most problematic cases, a doctor can remove sperm from the man's testicles and epididymides under local anesthesia, by needle puncture or minor surgery, then inject a sperm directly into an egg. Thanks to this technique, today even men whose testicles produce only a few extremely immature sperm can become fathers (see page 216).

In recent years some genetic causes of male infertility have also been found, particularly defects in the Y chromosome. Today the question of what happens if sperm with this type of defect are used to fertilize eggs by injection is quite controversial. Is there a risk of passing down the defect to the next generation, at least if the baby is a boy? Any clear answer to this question remains elusive.

## The woman undergoes more thorough testing

While male fertility testing can usually be carried out quite simply, on the basis of one or two sperm samples, more extensive testing must be done on the woman, who must not only become pregnant but also carry the baby to term. Ultrasound can give the doctor a good idea about the state of the various organs in her pelvic area. It can reveal possible ovarian cysts, or possible fibroids in the uterine lining, and show whether an infection might have led to the Fallopian tubes becoming swollen and blocked. It can show whether the uterus appears normal. A contrast medium is injected through the vagina and the cervix, making it possible to examine via X-ray or ultrasound the inside of the uterus, the thickness of the mucous membranes, and the Fallopian tubes. The physician can also determine whether the passageways from the ovaries to the Fallopian tubes are open, necessary for natural fertilization.

Some hormonal tests are also important, especially to check that the levels of prolactin, the stress hormone, are not too high, and that the progesterone level shoots high enough after the predicted date of ovulation to make fertilization possible.

If mechanical and/or hormonal problems are not treated successfully either with surgery or hormone injections, many women turn to in vitro fertilization (IVF) treatments, also known as test tube fertilization.

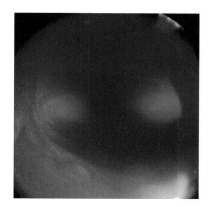

**Checking the uterus**

Using an endoscope and inserting it through the vagina and the cervix, a doctor may obtain a clear picture of the relatively small cavity that will be the home of the fetus during its many months of development (above). The openings of the Fallopian tubes, through which the fertilized egg normally enters the uterus, are also visible. Sometimes inward bulging fibroids (myomas) reduce the space in the uterus. This may necessitate medication or surgery prior to test tube fertilization.

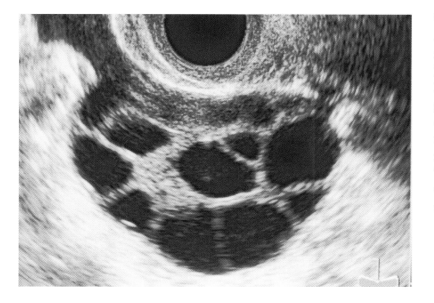

**Oocytes in the ovary**

As part of the process of test tube fertilization, the woman receives hormonal injections, after which a large number of oocytes begin to grow larger in her ovaries. Each one contains an egg that is viable for fertilization. An ultrasound scan indicates the number of oocytes. Vaginal ultrasound is also a guide when the eggs are gathered from the ovaries (left).

## Stimulating oocyte growth

As IVF treatment begins, the woman's own hormone production is suppressed, as medication turns off the function of her pituitary gland. She will then have daily hormone treatments of hCG (human chorionic gonadotropin) or clomiphene, most commonly for a couple of weeks, to stimulate a large number of oocytes to develop in her ovaries. If clomiphene has been used, hCG is often given as a final injection.

Just before the time when natural ovulation would have occurred, a thin needle is inserted, with the help of vaginal ultrasound, through the wall of the vagina into an ovary, where it extracts oocytes. The eggs are then examined, using a microscope, and placed in a special nutrient solution. They are kept in an incubator at 37.5°C (99.7°F). The ideal number of eggs to be harvested is eight to ten, but regulating hormone stimulation so exactly is difficult. Sometimes there will be too few mature eggs; at other times there may be far too many. This kind of overstimulation may cause abdominal pain and swelling, and the woman may need to stay in the hospital for a few days.

**A daily hormone fix**

Most women who have IVF treatment receive daily injections of pituitary hormone for about two weeks before the test tube fertilization. When possible, the couple manage the injections themselves. The hormone levels are adjusted so that a suitable number of follicles containing oocytes will begin to grow larger, and when the largest one is about 2 cm./0.8 in. across, a final injection of another hormone, hCG, is given.

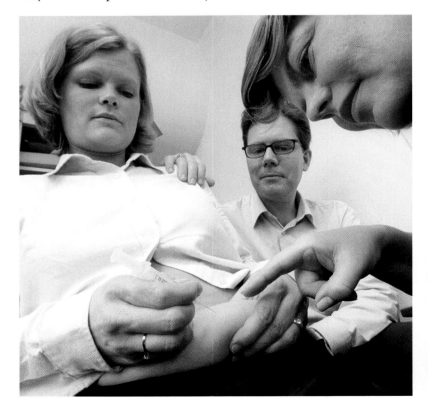

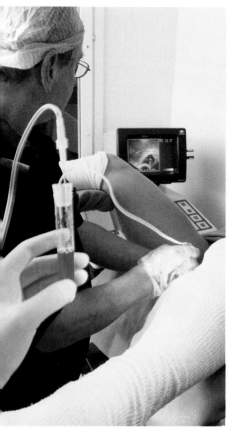

## Eggs suctioned off

Thirty-six hours after the woman has received her last hormone injection, the time has come to remove the eggs by suction, using a needle inserted through the vagina. Ultrasound helps guide the needle in the right direction.

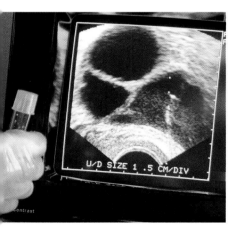

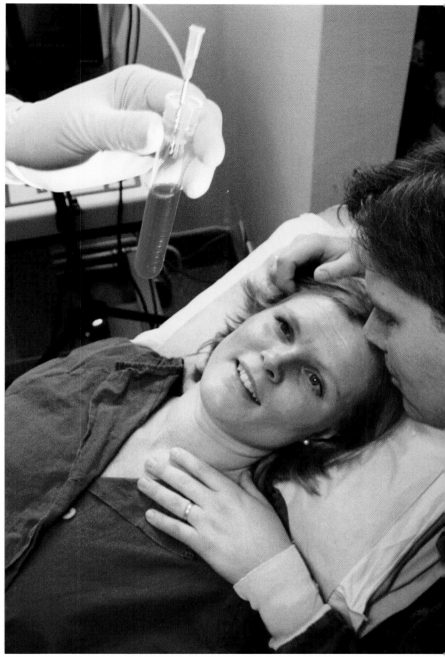

## Anticipation

The man is often present while the eggs are removed, and the couple is kept well informed as to the number of eggs found when the fluid has been removed. Each egg is a potential pregnancy. Often about ten eggs are found.

## Capturing the egg

The fluid removed by suction from each follicle contains not only the egg itself but also thousands of nutrient cells and possibly some blood. Experienced eyes using a good microscope can usually identify the egg very quickly. Each egg is removed, is transferred to a drop of nutrient solution, and is then carefully deposited in an incubator with a temperature of 37.5°C/99.5°F.

## *The egg and sperm meet—outside the body*

In conventional IVF fresh sperm, which the man has presented to the laboratory and which are then specially prepared, are introduced into this egg sample. In nature as few as a couple of hundred sperm may reach the egg, but in IVF it is possible to be far more generous. Often many thousand sperm are introduced to the culture dish, in order to maximize the chance of fertilization. After this encounter between sperm and eggs, usually lasting some sixteen to eighteen hours, the culture is examined, and it is easy to see microscopically whether fertilization has taken place.

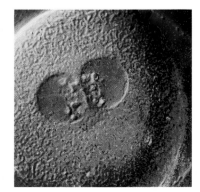

## Fertilization

Using a pipette, a large number of fresh sperm are added to some ten eggs, each of which is in its own little drop of nutrient solution (opposite). The miraculous transformation then takes place in the dim incubator: egg and sperm fuse, creating new life.

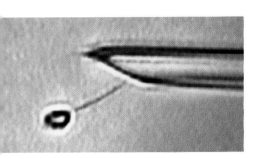

**First row seat**

With a narrow glass pipette, one sperm is sucked up out of a nutrient solution (above). Now the decisive moment has arrived. On the screen the couple can follow the process that may result in the baby they are longing to have (below). The selected egg is in a culture dish, being held in place with a thicker glass pipette and micro-manipulators connected to a microscope.

## Sometimes a sperm is injected right into an egg

In recent years a new technique has begun to be used, particularly when there are problems with the sperm. This alternative to conventional test tube fertilization is known as intracytoplasmic sperm injections (ICSI). In ICSI a single sperm is placed at the very center of the cytoplasm of the egg, using a thin glass needle. Even living sperm that are incapable of propelling themselves can fertilize an egg in this way. Even "immature" sperm, removed directly from the testicles and the epididymides, are able to fertilize an egg with the help of this technique.

Some are concerned that this new technology, which transgresses certain natural barriers, could imply risks for the unborn child. In response, some researchers have proposed that all embryos created in the laboratory should undergo genetic testing before being returned into the mother's body. Fertilization outside the body would then probably mean a far lower risk of genetic disorders than natural fertilization. Such a possibility is both exciting and potentially worrisome.

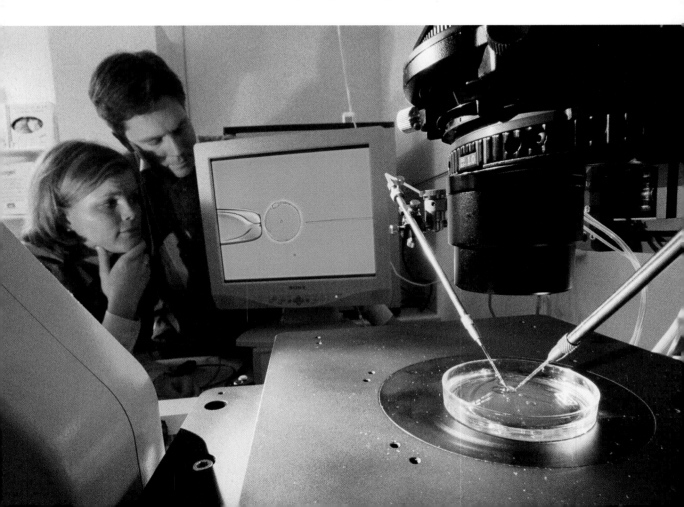

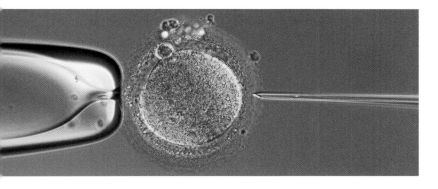

## ICSI-technology portrayed

Gently the pipette punctures the eggshell, and the sperm is injected into the cytoplasm of the egg, where the female nucleus is waiting. As the left-hand pipette holds the egg in place, the thinner pipette on the right is maneuvered into exactly the right position: this precision work requires years of practice to master. After the injection the egg quickly regains its rounded shape. In the photo below the contour of the head of the sperm can be seen in the middle of the oocyte.

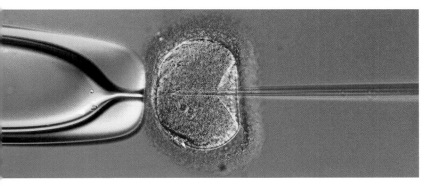

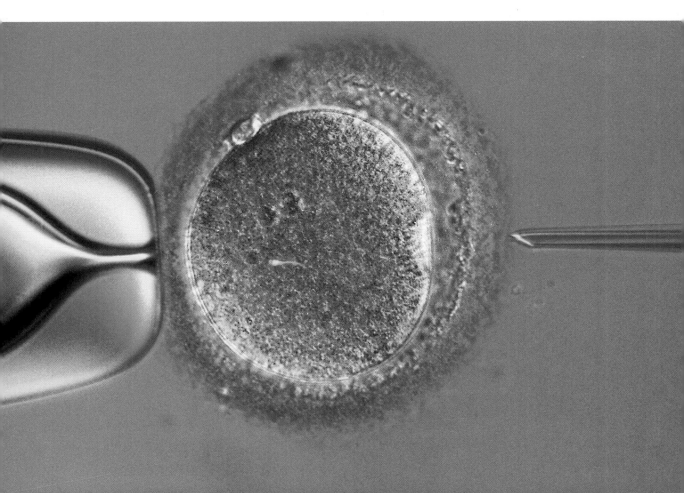

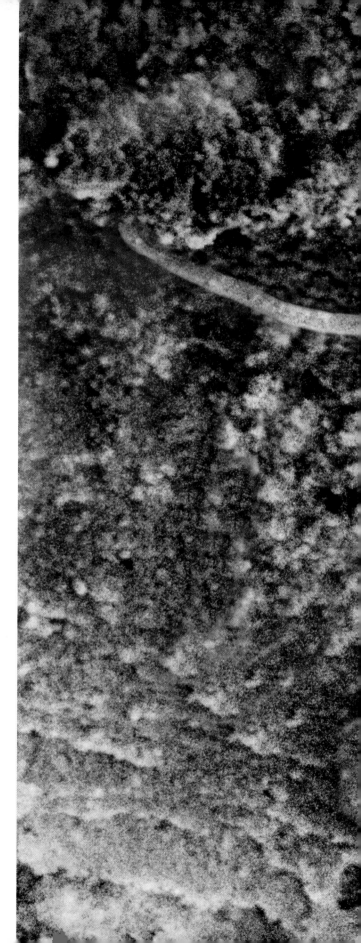

### Reaching the finish line

The chosen sperm has just been injected through both the eggshell and the cell membrane—the hole from the pipette is still clearly visible. Deep inside the cytoplasm the sperm can now complete the task of fertilization unassisted. It loses its tail, and the tightly packed bundle of genetic information in the head expands slightly in anticipation of its encounter with the nearby kernel.

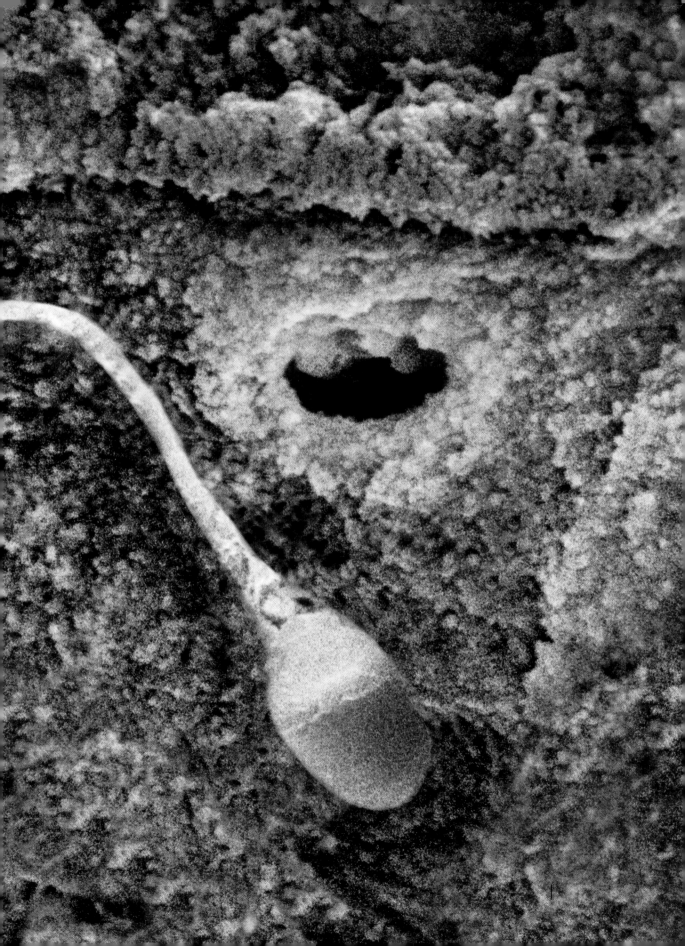

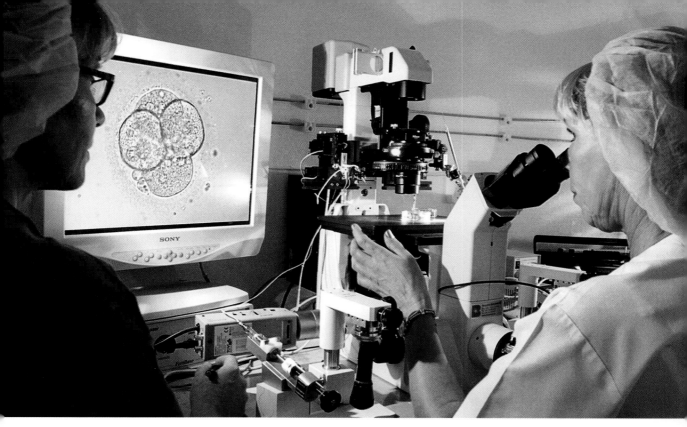

## Making the right choice

Two to three days after fertilization, as many as six to eight fertilized eggs may have begun to divide. One or two are selected with great care and are returned to the mother. The remainder are frozen for possible future use.

## Back to mother's body

When fertilization has taken place (with the aid of either conventional technology or the ICSI method), two distinct nuclei appear in the cytoplasm of the egg, one from the head of the sperm, which has become enlarged, and one that contains the genetic material from the egg. A few hours later, still in cell culture, the two nuclei fuse into a unique new genetic code, after which the fertilized egg begins to divide at twelve -to fifteen-hour intervals.

After two to three days the little embryo consists of four to eight cells and can be returned to the woman's body. Currently, fertilized eggs are often cultured for another two or three days, until they reach the blastocyst stage, before being returned to the womb using a thin plastic catheter carefully introduced via the cervical canal. All parties concerned now cross their fingers and hope that the embryo is healthy and able to implant in the womb. A week or two later a very sensitive hormonal blood test can be used to determine whether the woman has become pregnant.

Under normal circumstances only one egg per month (or in occasional cases two) is discharged from one of the ovaries. In IVF, with hormone stimulation, as many as ten eggs are obtained from every treatment. Six or seven of them may be fertilized and develop into embryos. If all these fertilized eggs were returned to the uterus at once, the result could be sextuplets or septuplets. Multiple births imply increased risks for both mother and children. Thus today, depending on the couple's feelings about selective reduction in the womb, only one or two embryos are usually returned after a treatment, while the others are frozen for possible later use.

In Sweden there is an explicit ambition to bring the number of twin births by IVF down to approximately three percent, the level nature, too, maintains, and to have no births at all of triplets and quadruplets after IVF.

The technology used to freeze embryos, as well as eggs and sperm, is under continual development. Legislators in many countries are concerned that lengthy storage of embryos in frozen form may lead to ethical and legal problems. Legislation on this issue varies greatly from country to country.

**Frozen life**

At first freezing eggs, sperm, and embryos may seem quite strange, but in reality this is an effective method of giving a woman more than one chance of becoming pregnant after hormonal stimulation.

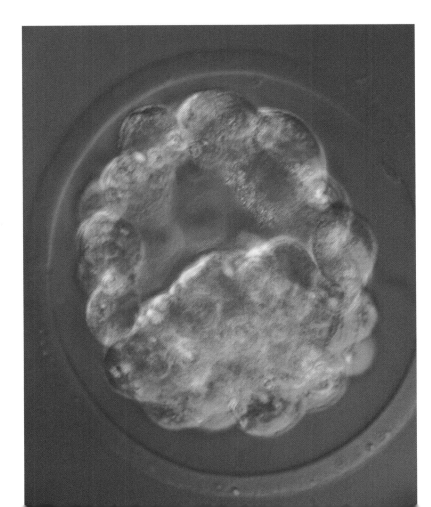

**Future perspectives**

Today improvements in culturing technology have made it possible to culture the fertilized eggs for four to five days before returning them to the uterus. This makes it much easier to determine which fertilized egg has the greatest chance of developing normally. Genetic analysis may, in the future, be the best sorting instrument and help prevent certain genetic disorders that could cause great suffering.

**Back to where it all began**

When the result of the joint efforts visits the hospital, there is great jubilation. When IVF techniques are used to help a couple become pregnant, teamwork is very much in evidence. The parents and the medical care personnel often become close.

*Every baby is a miracle, and creating a baby in a dish is miraculous indeed.*

## More than a million test tube babies

Nineteen seventy-eight, when the first test tube baby was born, was a historic year in the treatment of infertility. The baby was born in England as a result of the work of researchers Robert Edwards and Patrick Steptoe. Gradually the method came to be used to treat almost all kinds of female infertility. In 1992 the microinjection of a single sperm into the egg came into use, replacing virtually all previous treatment regimes for male infertility. Today more than a million babies have been born after fertilization outside the body, and major studies have so far shown that these children develop normally and that the risks for the child are no greater than the risks associated with natural fertilization.

Today most kinds of infertility can be treated successfully. But the best methods are extremely expensive, making infertility more a socioeconomic problem than a medical-technical one. Moreover IVF treatment is a great strain, both physically and psychologically, particularly for the woman involved.

Many countries are seeing an ongoing public debate as to whether in a modern, resource-intensive society people have a self-evident right to be helped to have children. Other issues in the debate include possible risks of infertility treatments for the woman's health in the long term, or for that of the baby. Is infertility an illness or the consequence of an illness? Ethical or religious considerations are also much debated.

Many questions are as yet unanswered. But infertility can cause great suffering for couples, both in countries struggling with uncontrolled population growth and in countries where birth rates are falling and the population is decreasing. IVF is one of a number of solutions to the complex problem surrounding childlessness. Adoption is another, and some people combine the two methods. Taking in foster children is still another way of having a family relationship with children and of experiencing the tremendous pleasure of playing an important role in the life of another human being.

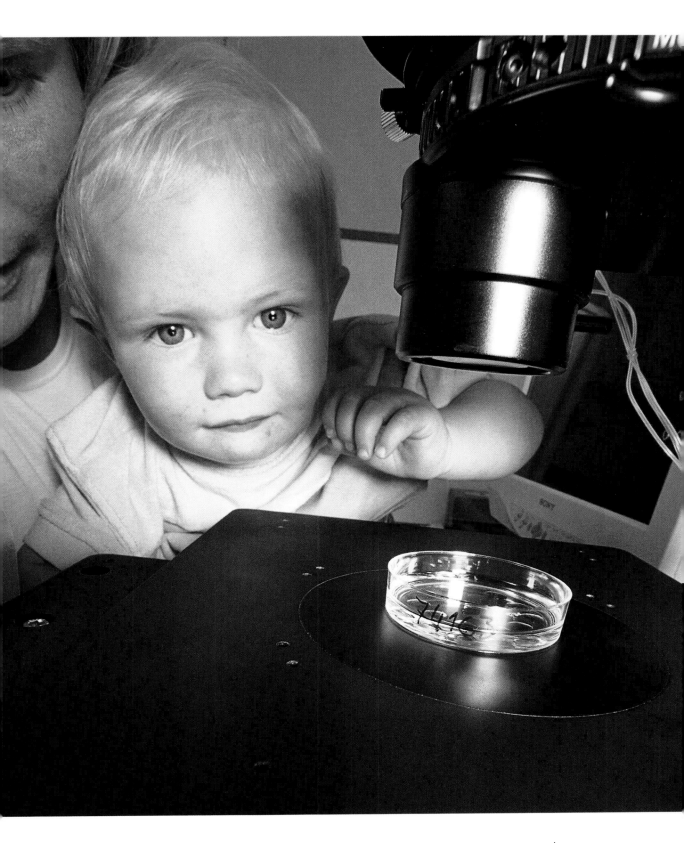

## Twelve years between pictures

Not everyone has a picture of himself just a few days after being conceived. Jonas does, and he is just one of the many children in the world who have come into being thanks to IVF treatments. Amazingly, the little cluster of cells on the screen has become a healthy boy approaching adolescence. IVF children themselves sometimes find it difficult to understand the process. On the other hand, since young children have no preconceptions about reproduction, they tend to be able to take in and accept the information they receive.

### "You were in a dish like that"

More than a million children have come into the world thanks to Robert Edwards, the professor and researcher who developed the successful IVF method.

### Different ways of becoming a family

IVF treatment has provided a chance for many infertile couples to make their dream of having a family come true. Jonas's parents tried in vain to have children for a long time—today they have three. They chose to adopt a son and had two children after IVF treatment. The arrival of each one made them happy and excited.

# The Adventure Begins

**Two have become three. A newborn discovers the world, while life shifts into a new gear for the parents, a time of special experiences and changing priorities.**

## Coming home

Arriving home for the first time with a new baby is a fantastic experience. The whole world looks different, feels almost unreal. Suddenly there is a new person constantly wanting love and attention. For the first few days everything focuses on the child, and time seems to stand still. Caring for a newborn is both a great pleasure and a huge strain, particularly the first time around.

Nursing and practical caretaking seem to fill every hour of the day and then some, and most first-time parents feel awkward and insecure. Why is the baby crying? Could I be doing something wrong? Many mothers are often quite exhausted, so the help and support of the father is particularly important. Nights are broken up with feedings until the baby begins to sleep through, which usually takes at least a couple of months. The baby then tends to sleep less during the day and shows more and more personality. He or she suddenly has habits and personal needs, as well as a will that may be strong indeed. It takes some time for the family to re-establish everyday routines and rhythms.

Six to eight weeks after delivery, the family makes a final visit to the obstetrician, to be sure that the mother has completely recovered from the physical aspects of childbearing. The parents also have a chance to review the delivery, and to discuss their choice of birth control during the nursing period and afterward.

For the baby, visits to a pediatrician or a well-baby clinic are vitally important, more frequently in the early months, and once a year in late childhood. Such visits are an important source of support, and they help the parents solve any problems that might arise. Most pediatricians have regular call hours, and parents should keep emergency numbers for the doctor, hospital, and poison center available at all times. Parents' confidence in their own judgment and competence will show day by day.

**The vital mother's milk**

For the first few months of a baby's life breast-feeding is best, and many women go on nursing for much longer than that. The American Academy of Pediatrics recommends one year. If, for some reason, the mother is unwilling or unable to breast-feed her baby, the alternative is infant formula.

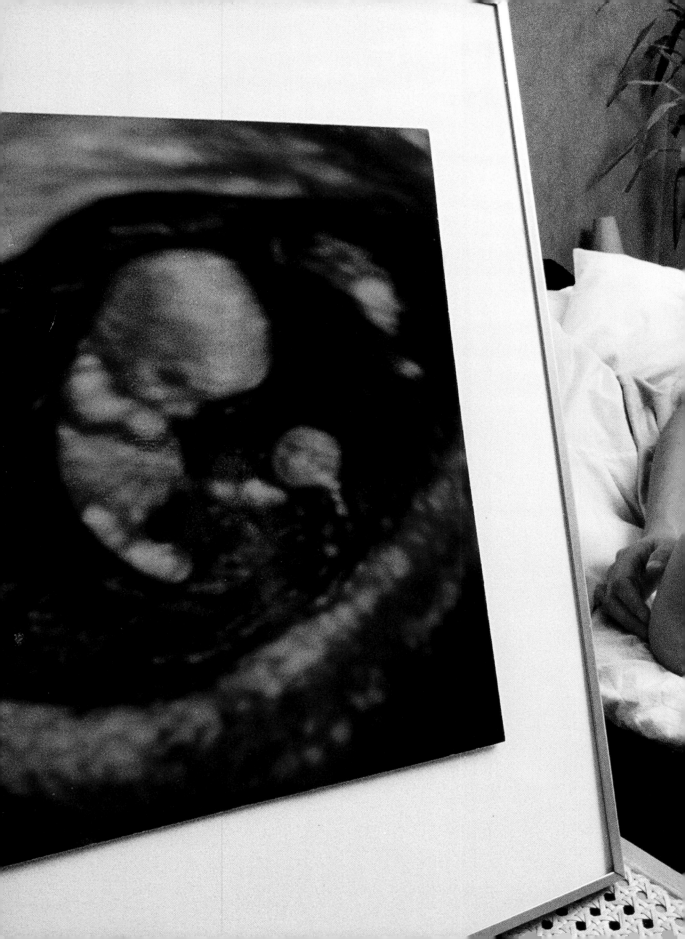

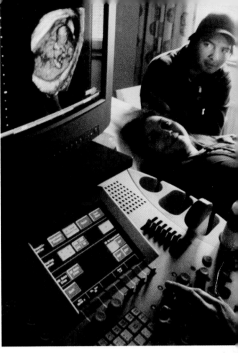

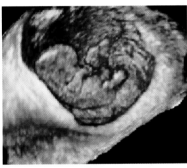

**How could she be so sweet?**

This father got to see his daughter tumbling around inside her mother in three-dimensional ultrasound as early as week 8 of pregnancy. Now here she is, sitting beside him in bed at home. The same baby—and yet so different.

*Once a picture flickering on a screen, she is now really a baby, full of energy.*

**Playing and cuddling**

A kiss on the tummy and a tickle between diaper changing and
feeding time—early infancy is a time of tenderness, closeness,
and perpetual physical contact. In these first months crucial
bonds are established, and a basis of security and trust between
child and parents is formed.

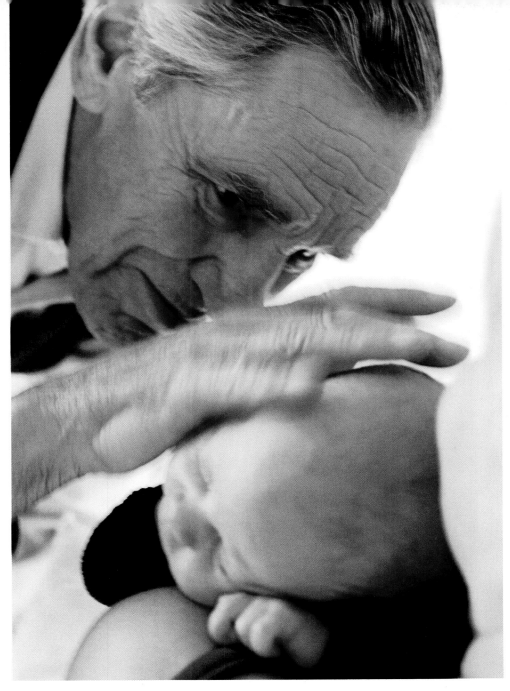

**Oldest and youngest in the family**

A child who comes into the world is always part of a larger context. The child is more than just the parent's baby—he or she may also be a grandchild or even a great-grandchild, a sibling, and a cousin. This great-grandfather on the baby's father's side is delighted with the new little branch on the family tree. The family heritage continues.

## About the photographs

Lennart Nilsson has been documenting the beginnings of human life for more than fifty years. His tireless efforts to portray the story of how every single one of us comes into the world has kept him at the cutting edge of scientific research. The photographic techniques and instruments he has used over the years span a wide spectrum of devices developed by the leading manufacturers of electronic and optical equipment.

This edition of *A Child Is Born* is composed primarily of new photographs, but it also contains some that are now classics. The work could never have been completed without close collaboration with researchers and clinicians, particularly gynecologists. Basic and clinical research are regulated by the ethics committees of universities, and their rules determine what can and cannot be done. Endoscopy and ultrasound technology have been used to capture many of the in utero pictures in this book. Other pictures were taken in conjunction with the in vitro fertilization process, with ectopic pregnancies, and with miscarriages. There are, however, no photographs of embryos or fetuses aborted either by medical or surgical means.

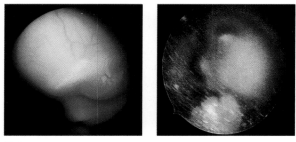

### Fetoscopy

It has been possible for over thirty years to use an endoscopic or fetoscopic instrument for surgical procedures inside the amniotic sac. This technique is used only when ultrasound does not provide enough information and the physician deems it necessary to examine the fetus further. The first "portrait" of a living fetus was taken by Lennart Nilsson at just such an examination (top left).

Today this technique is still utilized only when very stringent medical indications apply. Contemporary instruments, complete with lens and sheath, are smaller than one millimeter in diameter—no larger than an ordinary injection needle. Without increasing the risk of complications, this technique provides a relatively sharp image of a limited section of the fetus (top right).

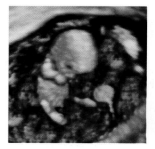

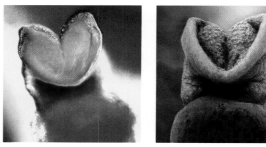

### Ultrasound

Ultrasound technology has recently made great advances, and today an ultrasound image can be interpreted even by a non-specialist. At specialist clinics three-dimensional ultrasound technology is now in use. This technique makes it possible to capture on the ultrasound monitor both very small embryos and larger fetuses.

### Special optical techniques

Some of the photographs in this book were taken using instruments with extremely wide-angle lenses, specially commissioned by Lennart Nilsson in order to be able to capture the entire fetus in the amniotic sac. These unique photographs could not have been taken using endoscopic or ultrasound technologies.

### Microscopy

The photograph above left was taken using light microscopy, which gives detailed, full-color images with excellent focus. The same image (right) has been captured using a scanning electron microscope, which enables enlargements to hundreds of thousands of times the natural size of the subject. Such images give a three-dimensional impression and, more important, a greater sense of depth than photographs taken using light microscopy. These pictures are developed in black and white and then tinted, in order to make them easier to comprehend, with a technique based on a translation of the gray scale to color. Some of the pictures taken with the scanning electron microscope have been digitally tinted.

# Acknowledgments

This book could never have come into being without all the marvelous couples who permitted me to share in and photograph their pregnancies and to be present when their children were born. I thank all of them from the bottom of my heart. My deep gratitude also goes to the staff at the hospitals and prenatal clinics, where I have worked to bring this project to completion; they have been of great assistance in so many ways. I would also like to thank technical engineer Jan-Åke Andersson, Assistant Professor Ursula Bentin Ley, Assistant Professor Elisabeth Blennow, Professor Hugo Lagercrantz, Professor Outi Houvatta, laboratory engineer Eija Matilainen, ultrasound technician Thomas Kratochwil, Assistant Professor Karl Gösta Nygren and Dr. Anita Sjögren. Without their great generosity, I could never have taken many of the photographs in this book.

I wish particularly to thank my wife, Catharina, for her support and assistance, and Lars Hamberger for his deep commitment to and engaged participation in this project. I would also like to thank the staff of Albert Bonniers Förlag who were the engine of this project for many years: Cecilia Bengtsson, Birgitta Emilsson, Susanna Eriksson Lundqvist, Robert Hedberg and Per Wivall, as well as my assistants Anne Fjellström, Anna Malmberg and Camilla Wodelius.

*Lennart Nilsson*

Many other specialists have also shared their knowledge, and I would like to thank them all warmly:

Professor Anne Grete Byskov, *Copenhagen University*, Cop.
Senior physician Lars Bäcklund, *Sabbatsberg Hospital*, Stockholm
Professor Robert Edwards, *University of Cambridge*, Cambridge
Professor Sturla Eiknes, *Trondheim University*, Trondheim
Professor Wilfred Feichtinger, *Wunschbabyzentrum*, Vienna
Professor Axel Ingelman-Sundberg, *Sabbatsbergs sjukhus*, Stockholm
Professor Elisabeth Johannisson, *WHO Geneva*, Switzerland
Dr. Chr. Kindermann, *Wunschbabyzentrum*, Vienna
Professor David de Kretser, *Monash University*, Melbourne
Professor Andreas Lee, *Allgemeines Krankenhaus*, Vienna
Professor Svend Lindenberg, *Copenhagen University*, Cop.
Dr. Michaela Munkel, *Donauspital*, Vienna
Professor Erik Odeblad, *Umeå University*, Umeå
Professor Leif Plöen, *Swedish University of Agricultural Sciences*, Uppsala
Dr. Henrik Rabeus, *Läkarhuset Odenplan*, Stockholm
Photographer Maud Reindal, *Saint Erik's Eye Hospital*, Stockholm
Senior physician Antal Szabolcs, *Södertälje Hospital*, Södertälje

**Danderyd Hospital, Stockholm**
Dr. Bengt Sandstedt
*Neonatal Department*
*OB-GYN and maternity wards*

**Göteborg University, Göteborg**
Secretary Eva Allén-Frizell
Professor Gunnar Bergström
Assistant Professor Erling Ekerhovd
Assistant Professor Charles Hanson
Assistant Professor Thorir Hardarson
Professor Anders Norström
Assistant Professor Bo Sultan

**Huddinge University Hospital, Stockholm**
Principal midwife Gisela Bergström
Senior physician George Evaldson
Dr. Ebba Hedin-Blomqvist
Professor Åke Seiger
Professor Magnus Westgren
*Center for Fetal Medicine*
*OB-GYN and maternity wards*

**Karolinska Institute, Stockholm**
Dr. Lars Brandén
Predoctoral fellow Kerstin Holmberg
Professor Bo Lambert
Assistant Professor Stefan Nilsson
Lab. assistant Berit Olsson

**Karolinska Hospital, Stockholm**
Radiological nurse Åsa Avango
Assistant Professor Oddvar Bakos
Dr. Marco Bartocki
Assistant Professor Olle Björk
Professor Marc Bygdeman
Dr. Björn Ekman
Senior physician Thröstur Finnbogason
Professor Seth Granberg
Principal midwife Gun Hermann-Jonasson
Senior physician The-Hung Bui
Lab. engineer Marita Johansson

Assistant Professor Ulrik Kvist
Lab. assistant Ann-Marie Lundberg
Midwife Marie Lönn
Dept. senior physician Lena Marions
Head nurse Inga Nilsson
Professor Bo von Schoultz
Senior physician Claes Silferswärd
Midwife Marie Strömberg
Section supervisor Inger Söderlund
Assistant Professor Håkan Wramsby
*OB-GYN and maternity wards*
*Pediatric X ray unit*
*Neonatal Department*
*Surgery wards*

**Uppsala University**
Princ. research engineer Leif Ljung
Research eng. Marianne Ljungqvist
Lab. engineer Tapio Nikkilä

**Sophiahemmet, Stockholm**
Assistant Professor Arthur Aanesen
Assistant Professor Rune Eliasson
Midwife Anna Franzon
Assistant professor Claes Gottlieb
Midwife Kristina Haglund
Lab. assistant Kaija Hyvönen-Töcksberg
Laboratory assistant Björn Loftås
Midwife Cecilia Lärksäter
Assistant Professor Lars Marsk
Assistant Professor Lars Nylund
Lab. assistant Ann-Marie Thörnblad
Midwife Margareta Stefenson
Biomedical analyst Eva Örn

*Ekens Prenatal Clinic, Stockholm*
*Östermalms and Gärdets Private Prenatal Clinic, Stockholm*
*Rinkeby Prenatal Clinic, Stockholm*
*Skanstulls Prenatal Clinic, Stockholm*

*Visby Prenatal Clinic*, Visby
*Hälsopoolen, Rosenlund hospital*, Stockholm
*SatyanandaYogacenter*, Stockholm
*McDonald's*, Stockholm
*Restaurant Martini*, Stockholm
*Råda Gästgifveri*, Mölnlycke

**My gratitude, for technical equipment and assistance to**
Professor Klaus Biedermann
Technician Åke Brunkener
Dr. Björn Ekman
Chemical technician Björn Holmstedt
B.Sc. in Medicine Torben Thölix
Professor K. Tanaka, *University of Tottori*, Japan
*Bertarelli Foundation*, Switzerland
*Flir Systems AB*, Stockholm
*General Electric—Kretz Technik*, Austria
*Viktor Hasselblad AB*, Göteborg
*Jeol*, Japan
*Karl Storz GmbH*, Germany
*Lorentzen Instrument AB*, Stockholm
*Nikon*, Japan
*Siemens*, Germany
*Zeiss*, Germany

**Special thanks for checking the facts to:**
Principal midwife Gudrun Abascal, *BB Stockholm*
Director Eivor Björkman, *Ekens Prenatal Clinic*
Dr. Richard I. Feinbloom, *Southern California Kaiser Permanente*
Professor Urban Lendahl, *Karolinska Institute*
Professor Lars-Åke Mattson, *Göteborg University*

# Index

Lennart Nilsson began his career in the 1940s. His brilliant photo essays have appeared in the leading news-magazines of the world, portraying kings and poets, slums and palaces. His greatest achievement is his visual exploration of the human body and the wonders of reproduction. His internationally bestselling books, *Behold Man* and *The Body Victorious*, culminated in *A Child Is Born*, which has sold tens of millions of copies worldwide in its many editions and has been translated into 20 languages.

Lars Hamberger, M.D., Professor and Chairman of the Department of Obstetrics and Gynecology at Gothenburg University, in Sweden, is known internationally for his research and writing in gynecology and fertility.

A Merloyd Lawrence/Delta Book

Publishing history
Delacorte hardcover edition published September 2003
Delta trade paperback edition / October 2004

Published by Bantam Dell
A Division of Random House, Inc.
New York, New York

Originally published in Sweden under the title *Ett barn blir till* by Albert Bonniers Förlag, Stockholm
Editors Cecilia Bengtsson and Susanna Eriksson Lundqvist
Designer Birgitta Emilsson
Production manager Robert Hedberg
Color drawings Åke Ahlberg (pp.18-19, 26, 33, 80-81)
B/w drawings Anders Palmgren
Coloring of scanning electron photos Gillis Häägg and Intermezzo Grafik (pp.27 above, 29 and 93)

Library of Congress Catalog Card Number 2003043854

ISBN: 978-0-385-33755-7
ISBN-10: 0-385-33755-8

Printed in the United States of America
Published simultaneously in Canada

4 5 6 7 8 9 0